# Authenticity and Urbanism

This book investigates authenticity, an important but controversial cultural concept, and its relevance for urban socio-spatial processes. Defining authenticity as an evaluative relationship with an origin, it unpacks a range of origins from a critical, non-essentialist perspective: self, author, substance, event, time, place, and society.

Authenticity is a key concept in contemporary culture. It was selected as the word of the year 2023 by Merriam-Webster, it is significant for UNESCO in designating the world heritage, a philosophical battleground, and reflecting concerns about truth and trust in the digital age. This book offers a clear understanding of the concept and discusses its implications for urbanism, understood as both material and relational, both physical and social. It critically examines whether authentic urbanism may be possible through seeking a solid ground in the self, in the act of making, in the substance of objects, in transformative events, in particular times and places, and in society and culture.

Analysing the complexities, nuances, and controversies of the concept of authenticity, and their implications for urban space and society, this book explores the promises and pitfalls of authenticity, looking for ways of avoiding its dangers and searching for its creative potential. Students and scholars in urban studies, geography, urban design, planning, and architecture, as well as in sociology and cultural studies, would find this interface between ideas and practices informative and insightful.

**Ali Madanipour** is Professor of Urban Design in the School of Architecture, Planning and Landscape at Newcastle University, UK.

# Authenticity and Urbanism

## A Critical Inquiry

Ali Madanipour

Routledge
Taylor & Francis Group
LONDON AND NEW YORK

Designed cover image: Getty images

First published 2025
by Routledge
4 Park Square, Milton Park, Abingdon, Oxon OX14 4RN

and by Routledge
605 Third Avenue, New York, NY 10158

*Routledge is an imprint of the Taylor & Francis Group, an informa business*



*British Library Cataloguing-in-Publication Data*
A catalogue record for this book is available from the British Library

ISBN: 978-1-032-58855-1 (hbk)
ISBN: 978-1-032-58856-8 (pbk)
ISBN: 978-1-003-45177-8 (ebk)

DOI: 10.4324/9781003451778

Typeset in Sabon
By codeMantra

# Contents

# Figures

*All photos taken by the author*

# 1 Introduction

Authenticity is a key theme in contemporary culture, to the extent that Merriam-Webster dictionary selected it as the word of the year 2023. It has found a growing allure in the context of globalization, intensified social diversity and inequality, heightened mobility, spread of digital technologies, accelerated temporality, and environmental decline. It is used in advertising, tourism, heritage, politics, environmentalism, and the arts, and has long been a battleground between major philosophical schools. While some see it as no more than a justification for narcissism and self-seeking behaviour, or a promotional tool to sell products and turn forms of life into presentable commodities, others see it as a necessary cultural concept and an important ethical ideal. Its wide appeal, its elusive connotations, and its controversial usage entail critically investigating its meaning and value for socio-spatial analysis. It is a quality that is often associated with individuals, objects, and practices, a yardstick to judge them as authentic or inauthentic. The question is whether this concept is meaningful or empty jargon, and if it means anything can it apply to cities, which are agglomerations of multiplicity and diversity? If cities can be or become authentic, what is the role of urban planning and design?

Authenticity is used in many contexts and with a variety of meanings, often taken for granted without attracting much attention. However, when it becomes the focus of attention, it reveals its elusiveness and its potential for controversy. When I discussed my work on authenticity at a conference of urban historians, my interlocutors seemed to consider the notion of authenticity as an essential ingredient of the heritage discourse, as widely used by the UNESCO in designating the World Heritage Sites. In this context, raising critical questions about authenticity seemed to be treated as unnecessary and unjustified. In contrast, when I discussed the concept at a conference of urban planners, it was treated with a degree of hesitation and suspicion, and even resistance by some who associated it with reactionary politics. Meanwhile, the 2024 annual conference of the Association of European Schools of Planning (AESOP) in Paris invited its participants to reflect 'on the authenticity and efficiency of traditional plans', so that new ways of planning could be found, implying the necessity of searching for more authentic and efficient plans.

DOI: 10.4324/9781003451778-1

The diversity of these approaches among urban researchers demonstrates a range of perspectives and temporalities, from a concern with the preservation of the past to the transformation of the future. More broadly, they exemplify the elusive and controversial nature of the concept of authenticity.

Authenticity has been the subject of interest in different fields. These fields include philosophy (Adorno, 1973; Berman, 2009; Bronner, 2021; Ferrara, 1998; Golomb, 1995; Taylor, 1991), environment (Dudley, 2011), politics (Kohl et al, 2021; Umbach and Humphrey, 2018), arts (Aldrich and Hackforth-Jones, 2012), social and cultural studies (Finney, 2019; Vannini and Williams, 2016; Williams and Schwarz, 2021), anthropology (Lindholm, 2007), ethnicity and race (Garratt, 2018), business studies (Beverland, 2009; George et al, 2017; Thacker, 2016), and popular psychology (Joseph, 2016; Seligman, 2002). There are also investigations into aspects of urbanism, particularly planning, heritage conservation, and tourism (Lovell and Bull, 2019; Lovell and Hitchmough, 2020; Tate and Shannon, 2019; Weiler and Gutschow, 2017). This book draws on this rich literature but develops a distinctive inquiry into the relationship between authenticity and urbanism through a critique of the origins on which the claims of authenticity are based. It is a critical, interdisciplinary inquiry into authenticity and urbanism, with a focus on the interface between theory and practice, drawing on philosophical ideas, social practices, political concerns, and spatial organizations.

Each chapter starts with a theoretical examination of a specific origin for authenticity, followed by exploring the implications of these theories for urban space and society. This approach combines the theoretical, as articulated in philosophy, humanities, and social sciences, and the practical, as developed in everyday life in urban space and practised in architecture, planning, and urban design. This is based on the notion that theoretical ideas do not constitute an abstract realm that is separate from concrete reality, but they are embedded in that reality (Aristotle, 1991). Therefore, ideas and practices are not detached from the world but are mediated through physical and social environments (Hegel, 1975), and while these contexts set the frameworks for action, they do not determine it (Badiou, 2005). The approach is based on the argument that to understand urban processes, we need a critical approach that combines political economy with cultural analysis (Lefebvre, 1991). In analysing authenticity, what is needed is a non-essentialist approach that examines its dangers and limitations while looking for its creative potential as a quest for truthfulness (Adorno, 2002; Benjamin, 1969; Foucault, 1986; Taylor, 1991). This investigation is part of a process of research into the inseparable social, spatial, and temporal dimensions of urbanism (Madanipour, 2017; 2020; 2021; 2023).

This book asks whether authenticity is a valuable concept, whether there is such a thing as an authentic city, and whether we can speak of authentic urbanism. While authenticity has been discussed in other domains, its relations with urban analysis have been less researched. The book's key questions are: is authenticity a meaningful concept? Is it a useful concept in the context of cities? Can cities be called authentic at all, and if so in what sense? What would an authentic urbanism be like? Is it a contradiction in terms, or can it be a motive for change? The book looks for answers in philosophical debates, in the digital, institutional, and physical domains, in everyday urban social life, and in the practices of shaping and managing cities. From a critical standpoint, the book searches for the meaning of the concept and its creative potential in the urban field.

## In search of an origin

The starting point of the inquiry has been the common meanings of authenticity, as articulated in major dictionaries. In selecting it as the word of the year 2023, the Merriam-Webster dictionary highlighted that the stories and conversations about this word, in relation to Artificial Intelligence, celebrity culture, identity, and social media, had led to a substantial rise in its use. In the same year, a Google search for the word 'authentic' showed 1.6 billion results in a fraction of a second, with the first page referring to the meanings of the word in different sources, as well as to subjects such as authentic beauty, authentic leadership, authentic happiness, authentic person, and authentic self. This first page also presented a separate box on 'authenticity' as a philosophical concept. A search for 'authenticity', meanwhile, showed even more results, exceeding 1.7 billion, with the first page being mainly about the meanings of the term, and how the subject is a concept of personality in psychology, psychotherapy, philosophy, and aesthetics. The selection of authentic as the word of the year and these large numbers of webpages that refer to authentic and authenticity are a witness to the wide appeal and extensive use of the term today.

The allure of the concept has varied across time and culture, oscillating between peaks of high interest and troughs of inattention. Google Books Ngram Viewer shows that in the English language, authentic and authenticity were highly popular words at the start of the nineteenth century, a period in which European romanticism flourished. However, this peak was followed by a steady decline throughout the century to its lowest point in the early twentieth century, when it started a gradual rise, leading to a sudden and sustained growth of popularity from the early 1980s up to now, although not yet to the level that had been reached more than two centuries ago. In the German language, a somewhat comparable pattern is observable, although here its recent usage has surpassed the early nineteenth century. In French language, however, it shows a lower profile for the nineteenth century and the first half of the twentieth, but an increasing popularity from the mid-twentieth century onwards, when it became the cornerstone of existentialism as a popular philosophical movement in France.

Despite selecting it as the word of the year, the Merriam-Webster dictionary admits that the word is hard to define and subject to debate. The Merriam-Webster online dictionary lists five meanings for authentic, including a mode in music and the obsolete sense of authoritative (merriam-webster.com/dictionary/authentic). According to the other three meanings, which are the most frequent senses of the word, it means 'worthy of acceptance or belief as conforming to or based on fact; conforming to an original so as to reproduce essential features; made or done the same way as an original'. It also means 'not false or imitation: real, actual', as well as 'true to one's own personality, spirit, or character'. Initially, there seems to be a surprising contradiction in these meanings, as they seem to refer simultaneously to an original as well as to a copy of an original, a reproduction that is not an imitation. This indicates the need to dig deeper.

As a noun, according to the Oxford English Dictionary (OED), the word 'authentic' refers to a mode of music and a collection of new laws published in the sixth century by the Roman emperor Justinian. But the main application of the word in common usage is as an adjective (OED: authentic). It affirms an expression to be 'in accordance with fact or stating the truth, and thus worthy of acceptance or belief; of established credit; able to be relied on; truthful, accurate'. The main meaning of the word is being 'Genuine; not feigned or false'. Therefore, for a document, artefact, or artwork, being authentic means 'having the stated or reputed origin, provenance, or creator; not a fake or forgery'. Something that is authentic 'is the actual thing or person; that rightly or properly bears the name; genuine, real'. It presents 'the characteristics of the original; accurately reproducing a model or prototype; made or done in the original or traditional way'. Being authentic means that it 'truly reflects one's inner feeling; not affected, unfeigned', and, in philosophical terms, 'a mode of existence which has its basis in self-awareness, critical reflection on one's goals and values, and responsibility for one's own actions; of or relating to this' (OED: authentic).

In different European languages, there are associations between authentic, author, and authority (see Chapter 3). In its broadest meaning, the word authentic describes a phenomenon itself or an accurate representation of it; the key question appears to be about the quality and accuracy of that description and representation. A fake representation is not reliable, as it pretends to be real, whereas an authentic representation seems to be explicit and reliable. Authenticity, in turn, means being real or genuine. It refers to 'The fact or quality of being true or in accordance with fact; veracity; correctness… accurate reflection of real life, verisimilitude' (OED: authenticity). In the case of objects and representations, such as documents, artworks, etc., it refers to the fact or quality of being authentic, being an accurate reflection of a model or exemplar, or being traditionally produced or presented. In the case of states of being, it indicates the quality of 'truthful correspondence between inner feelings and their outward expression; unaffectedness, sincerity', or a mode of existence based on self-awareness, critical reflection, and responsibility for one's actions. It also refers to the 'fact or quality of being real; actuality, reality' (OED: authenticity).

These meanings are reflected in authenticity's synonyms, which are listed as accurate, authoritative, convincing, credible, legitimate, original, pure, reliable, true, and trustworthy (Thesaurus: authenticity), all indicating the power of an origin, and the ability to provide evidence for that power's involvement, on which the judgement of authenticity relies. This is also reflected in its antonyms: corrupt, doubtful, false, implausible, improbable, incredible, invalid, unlikely, unreliable, untrustworthy, counterfeit, fake, falsified, unauthorized, ungenuine, unreal. Authenticity seems to offer the promise of truthfulness and trustworthiness, and its absence or opposite conditions are judged very harshly. In ordinary language, something is considered authentic 'when it proves itself unique, or is particularly trustworthy (e.g., "the real thing"), or meets high expectations (e.g., "genuine")' (Zuidervaart, 2007:84).

There are therefore many layers of meaning in the concept of authenticity, which need to be unpacked and discussed. These meanings are not merely terminological but also conceptual. At a glance, authenticity appears to trace a link between a representation and an event, between a situation and an origin or an originator, a correspondence between product and producer, between work and author, between act and actor. It appears to be an attempt to establish a causal link between a will and a state of affairs, while concerned about the quality and accuracy of that correspondence. Authenticity becomes the possibility of establishing believable correspondence between a state of affairs, a source of power, and a representation. It involves the questions of truth, truthfulness, and trust. It implies a search for some form of certainty, discovering a reliable foundation behind an ambiguous appearance. Authenticity is the expression of a judgement about the quality of something or someone. Investigating it raises a series of critical questions about the determination of that quality, of the ability and legitimacy of judges, and of the standards and procedures of judgement. Who decides something or someone to be authentic or inauthentic, on what basis, and for what purpose? Moreover, what are the origins on which a claim of authenticity is made? In this book, I examine seven such possible origins: self, author, substance, event, time, place, and society, trying to find out whether and how reliable they are as the platforms on which claims for authenticity can be made or evaluated, and how they might apply to urban life and space.

## A critique of origins

The book analyses different types of origin that have been used in the discourses of authenticity, each in a separate chapter, and the different cultural expressions that are associated with these origins, to find out their implications for urban life and social space. The titles and the subject matter of the chapters, therefore, indicate different possible origins which are used as the criteria for judging something to be authentic or inauthentic. Each chapter examines the concepts and their application to physical and social expressions that may be relevant for urban analysis.

After this introductory chapter, Chapter 2 explores the idea of the 'self' as the key reference point for authenticity. Drawing on insights from neuroscience and cognitive science to psychoanalysis, social psychology, and philosophy, it examines the difficulties of finding the self as a solid platform for authenticity. The presuppositions inherent in the idea of being true to the self are discussed, and the self is analysed as embodied, fragmented, imagined, and mediated, seeking social recognition, in a dialectical relationship with the environment. This finds spatial manifestation in the relations between the inside and the outside spaces, the range of private to public spaces that make up the human environment.

Chapter 3 examines the 'author' as the possible origin for authenticity. This is analysed through the role of authorship in the making of four domains: self, text, work of art, and city. This chapter problematizes placing the emphasis on the authority of detached individuals, genius authors, artists, or designers for these works, which in fact are produced by many actors at work, including the readers and viewers of texts and artworks, and the inhabitants of cities. Urban history and urban development processes show that cities do not have single authors but are the products of a complex set of agents, instruments, relations, and transformations.

Moving beyond the self and the author, the source of authenticity may be searched in the 'substance' of the objects, which is the focus of Chapter4. It examines what objects are made of and how, rather than who made them, a shift from author to object that critical theorists such as Benjamin and Adorno advocated as a critique of authoritarianism, opening aesthetic theory to social and political critique. From uniqueness and aura in artworks to the form and content of objects, their relation to society, and the role of technology in the production of the urban environment, this chapter analyses whether and how the quantity and quality of the works, and their methods of production, can contribute to a judgement of authenticity.

The origin of authenticity may be considered as 'events', unique occurrences that constitute the physical and social dimensions of the city, defying social conformity and its presumed inauthenticity. Chapter 5 analyses the city as a pattern of events, in the interplay of regularity and randomness, the formation of habits and routines in the city, and the significance of major events and everyday occurrences. The experiences of the city may be limited to surfaces, restricting the possibility of going beyond appearances, raising the question as to whether and how performances may be judged to be authentic. The epistemic dimension of authenticity in its relationship to truth and truthfulness is examined, especially in the accounts of events in a digital world.

Chapter 6 explores 'time' as a widely held origin of authenticity, which has led to two competing paradigms: a point of reference in the present, used as a platform for the construction of the future, and, more often, a point of time in the past as the source of authenticity in the present. In response to accelerated temporality, radical transformation of urban society, and the resulting interrupted identities, some sense of continuity is sought by making connections to a real or imagined point in time. This chapter discusses cultural heritage, especially the principles and practices of UNESCO World Heritage where authenticity is a central concern. It critically examines how narrative identities are constructed, and how memories are contested, as exemplified by the struggles over colonial monuments.

Chapter 7 examines 'place' as the point of reference for authenticity. Authenticity is analysed as a desire for specificity and particularity, for the unification of a divided self and society, through a return to a specific place, a return to the imagined purity and simplicity of home or the natural world. However, the person, society, and the elusive spatial origin have all changed and a return may not be possible or desirable. The connection to the place may trigger reactionary or emancipatory forces. Meanwhile, the idea of an authentic city is taken up by the claims of placemaking but often reduced to an image of vibrancy that may mask the practices of reproducing social hierarchy through placing and distancing.

In Chapter 8, 'society' and culture are investigated as the possible platforms for authenticity. In the context of diversity and complexity of society, the multiple forms of urban authenticity are discussed, and the values that are used in the judgement of authenticity are critically assessed. Can authenticity be achieved through democratic means, and what does it mean for a city to be true to itself? Authenticity can be a powerful critique of the social order, and therefore be adopted as an ethical ideal. However, it is argued that it is simultaneously an integral part of the social order, carving out a sphere of particularity that is in continual interplay with the forces of similarity and generality.

Chapter 9 brings together the key themes of the book, proposing a concept of authenticity, and re-examining the implications of this concept for urbanism in reference to the seven origins of self, author, substance, event, time, place, and society. Authenticity is based on a process of authentication, seeking an origin as a solid ground in the world of contingency. That solid ground is sought in the self, in the act of making, in the substance of objects, in transformative events, in a particular time and place, and in society and culture. The ground, however, is never unitary and solid, and authentication is always a process of selecting, bridging, and constructing, rather than treading along a paved path towards a well-signed, fixed destination.

# 2 Self

Authenticity has often been defined as being true to the self. Therefore, the first step in the investigation of authenticity takes us to the notions of self and personal identity, which is where most of the debate about authenticity takes place. This search takes us to different fields in arts and sciences that have investigated the self.

What is the self that is taken to be the origin of this authenticity, and what are the ways of reaching it? Is it a detached, fixed entity or is it formed dialogically and changed continually? What are the impacts of self-seeking beliefs and behaviour on society? Can a pure realm be found inside the body that constitutes that clearly defined self, or is the self a shifting pattern of perceptions, and a mere reflection of the social environment? From neuroscientists to psychoanalysts, critical theorists, phenomenologists, and poststructuralists, this self has been questioned from different perspectives. Should the search for authenticity be dismissed as dangerous and misguided, or is there a positive dimension to this demand? This chapter problematizes the concept of the authentic self and critically examines its implications for urban life, particularly in the relationship between subjectivity and behaviour, between the individual and society, between the inside and outside, between the public and private spaces. This chapter starts by analysing the notion of authenticity as being true to self and the presuppositions that are inherent in that notion. This is followed by looking for this self in the biological processes of the brain and the body, in the psychology of the divided self and the imaginary self, and in the social self, with the implications of the relationship between the inside and outside for space and society.

## Be true to yourself

The notion of authenticity in individuals has often been summarized in the maxim 'be true to yourself', which has been widely used in the popular culture as well as in the classical literature. A glance at the thousands of websites that mention this maxim shows the diversity and range of attempts at articulating this adage into philosophies of life and step by step guides on how to achieve it. In response to the pressures of mass society, where standardization

DOI: 10.4324/9781003451778-2

and routinization drive the patterns of everyday life, people are encouraged to discover who they are, what is the core of their identity, and live life according to that identity. This search for a true self has become a cornerstone of the modern culture, finding diverse manifestations in everyday discourses as well as in philosophical debates. Existentialist philosophy has advocated authenticity as its key concern, and for modern liberal societies, the notion of individual freedom and autonomy is a primary principle. The significance of authenticity has been widely emphasized in popular culture, especially in advertising and consumerism, which seem to encourage and provide a licence for excessive individualism and selfishness.

An analysis of this maxim suggests that it is based on the following five presuppositions: ontological split, differential value, epistemological possibility, normative desirability, and power availability. First, it presupposes an ontological split in you, which shows that you are made of two parts, whereby you and your self are not the same thing. It suggests that there is something called the self that is inside you but somehow different from you. You are not the same as your self, which implies that there is a discrepancy between who you seem to be and who you really are, a gap between the outside and inside, between the appearance and the essence of a person. That inner self constitutes your essence, which is different from your appearance. The second presupposition assigns different values to the two parts of this ontological split. It is based on the idea that the self, as the inner layer, has a higher value than the outer layer, and therefore it should be given priority. It is the original you, the more truthful you, your essence, and therefore it is more valuable. The second presupposition is closely related to the first, in that the idea of splitting something into two parts is directly related to evaluating the parts differently.

Third, it presupposes epistemological possibility, as the maxim suggests that you can know your inner self, which is the subject of the famous invitation to self-knowledge. It is assumed that the self is not readily known, and you are urged to contemplate on this self and get to know it. The maxim does not explain how to be true to yourself and how to know it, but it expresses no doubt that you are capable of doing so. Fourth, normative desirability, whereby being true to this original self would require establishing a clear correspondence with it, unifying your outer and inner layers, by bridging the gap or by removing or adjusting the outer layers to discover and display that inside essence. This is the core message of the maxim, whereby you are encouraged to make that connection, remove the curtain, and unify the essence and the appearance. The fifth presupposition holds that you have the power to do so, able to rearrange your life in a way that your inner self can be revealed, and the outer layers may be changed to coincide with that inner self. Based on these five presuppositions, a maxim is presented and repeated throughout the centuries as words of wisdom and as instructions for ethical life.

Each one of these presuppositions addresses one or more of the meanings of the term authenticity, all suggesting a relationship between a stable interior and an ephemeral exterior. In the first sense, where reference is made to the existence of an inner self, authenticity means purity and originality, and being true to yourself means acknowledging the existence of that inner source. In the second sense, where the inner self is valued higher than the outer appearance, to be authentic would imply having credibility and reliability. In the third sense of self-knowledge, uttered ever since Socrates and even long before him, authenticity would indicate the ability of getting access to and revealing the truth that lies behind the surface. In the fourth sense of the unification of inner and outer layers, authenticity would mean accuracy, credibility, and trustworthiness. In the fifth sense of the ability to become true to yourself, authenticity would mean having the agency and authority to change your own circumstances.

To be true to yourself, therefore, is a recipe for the unification of the inside and the outside, between mental processes and bodily behaviours, essences and appearances, private characteristics and public performances. These inherent presuppositions, however, cannot be taken for granted and are not as straightforward as they may appear. They need to be critically assessed, which leads to a range of questions: is there such a thing as a self, as distinctive from a person, and if so, what is it made of? Is this self the better part of a person? What are the values that enable such a judgement, and where do they come from? How can the self be known, and how is it possible to change the person to be unified with this inner self? How far is the unification of the inside and outside possible or desirable? What are the implications of being true to yourself for others, and if it becomes a reality, in what ways would urban society and space differ from now?

## The embodied self

Searching for the inner self takes us to the hidden realm of the mind, starting with scientific research into the brain as the physical and biological underpinning of the self. While the philosophers of mind have engaged with the subject for millennia, cognitive science and neuroscience are new disciplines that investigate the structure and function of the brain and neural networks, inquiring into longstanding questions about mind and consciousness. An influential view among these researchers is that the brain, in interaction with the rest of the body, produces the mind as a process; when mental processes come together in a centre of attention, the result is consciousness, and when this consciousness is directed inwards, the result is a sense of self (Bennett and Hacker, 2007a; Bermúdez, 2023; Chipman, 2014; Damasio, 2004; Dennett, 2013; Greenfield, 2000; Lau, 2022; Pfaff et al, 2022; Robinson, 2007). This account, however, has had to face questions about how research into brain activities can shed light on the contents of the mind and on psychological experiences. This may be discussed through at least two sets of overlapping challenges: how to explain subjectivity and the relationship between subjectivity and materiality.

The first challenge is explaining consciousness and human subjectivity, which is mainly understood through a first-person perspective. Cognitive science investigates the mind from a third-person perspective through the empirical methods of natural science, researching the processing of information that lies behind observable behaviours and capacities. It studies the brain, its structure, and its operations, with various methods such as brain imaging, reaction times, and verbal reports. It maps the brain and observes how its different regions function, learning takes place, and language is used. Core ideas of cognition have included symbolic representation and information processing, as well as neural networks and their connections. Alongside cognitive science, which focuses on cognition, affective science and emotional studies have investigated subjects such as fear and pain. Despite these advances, the hardest problem for cognitive science has been explaining the subjective experience (Bermúdez, 2023).

A prominent approach in cognitive science investigates the brain as an information processing machine, and therefore its growth as a discipline is closely related to the rapid development of computer science, machine learning, Artificial Intelligence, robotics, and big data (Bermúdez, 2023; Chipman, 2014). The information-processing model of consciousness has researched some forms of unconsciousness, through studying the impact of brain-damage on information processing. But it has been unable to account for the human experience, that is, the first-person perspective which is engaged in feelings, emotions, and meanings (Bermúdez, 2023). Cognitive neuroscientists have attempted to explain subjective experiences, but from a mechanistic perspective (Lau, 2022).

Some philosophers of mind have argued that the mind cannot be reduced to its physical components: the brain is 'a biological machine' that produces consciousness, and therefore it is 'a conscious machine', but consciousness essentially 'consists in inner, qualitative, subjective mental processes' (Searle, 1997:202–204). Consciousness is observer-dependent, with 'a first-person or subjective ontology', which 'cannot be reduced to anything that has third-person or objective ontology' (Searle, 1997:212). In response, other philosophers of mind have questioned the first-person account as untrustworthy: 'you *can* misremember, misinterpret, misdescribe your own most intimate experiences, covertly driven by some persuasive but unreliable bit of ideology' (Dennett, 2013:352). Dennett (2013:342) proposes the method of heterophenomenology, which is 'the study of first-person phenomena from the third-person point of view of objective science'. In doing so, it interprets the subjective phenomena as believed and performed by the subjects of investigation.

Cognitive scientists have developed two major theories on the functional role of consciousness. First, the higher-order thought theory, which is mainly rooted in philosophy, holds that consciousness is the result of a higher mental state that thinks about itself. This theory, however, is not always supported by empirical evidence, and some cognitive scientists question it as circular and mysterious (Bermúdez, 2023). Second, the global workspace theory, which has found more supporters in cognitive science and neuroscience, is closely related to attention. Attention functions as a gatekeeper that filters out unnecessary data and amplifies the relevant information by bringing it out from the nonconscious and into the workspace. Rather than a single location, this workspace is thought to be a distributed network of high-level processors that connect different regions of the brain (Bermúdez, 2023).

In the philosophical ideas that are close to cognitive science, the self becomes closely related to the cognitive scientists' favourite definition of consciousness as a workspace in which attention brings different sense data and mental processes together. According to philosopher Daniel Dennett (2013:334), the self is 'the same *kind* of thing as a center of gravity, an abstraction that is, in spite of its abstractness, tightly coupled to the physical world', and therefore it can be scientifically studied. A centre of gravity in any object is not a physical thing, but 'a theorist's fiction', which unifies and makes sense of a collection of processes that make up the object. Selves are different from centres of gravity, as they have trajectories in space and time, and 'they gather as they go' (p.336). These are temporal processes that include the past and the future, memories and plans, so 'Each self is a person, with a biography, a "backstory", and many ongoing projects' (p.336). The self as a narrative centre of gravity is 'a sort of live-in manager' that owns a body (p.337), but this abstract entity with no material existence does not directly cause anything.

The second challenge that cognitive researchers have grappled with is the relationship between subjectivity and materiality, in other words between mind and body. Writing in 1637, Descartes believed in the separation of mind and body as two different substances: 'this 'I', that is to say, the mind, by which I am what I am, is entirely distinct from the body' (Descartes, 1968:54). This was an influential formulation that reflected the ancient beliefs in the separation of body and soul, and while it is no longer a mainstream idea in science and philosophy, it continues to be widely held in popular beliefs. As an early critic of this Cartesian dualism, Spinoza held the body and mind to be two aspects of the same thing. According to Spinoza, the mind was formed of the knowledge of the body: 'The object of the idea constituting the human mind is the body', and 'the human mind is the idea itself, or the knowledge of the human body' (Spinoza, 1996, Book II, Proposition 13 and proof of Proposition 19).

For the neuroscientist Antonio Damasio, who sides with Spinoza's aspect-dualism against Descartes' substance-dualism, 'the mind is a process, not a thing'; it depends closely on the workings of the brain and 'arises from or in a brain situated within a body-proper with which it interacts' (Damasio, 2004:183, 190–191). The integrated organism is formed of a body and a brain, and the brain creates and manipulates 'mental images (ideas and thoughts) in the process we call mind' (p.194). Damasio (2004:184) distinguishes the mind from consciousness: 'consciousness is the process whereby a mind is imbued with a reference we call self, and is said to know of its own existence and of the existence of objects around it'. In this way, an interactive chain of relations is set up between body, brain, mind, consciousness, and self.

The relationship between body and mind is clearly evidenced in the ways in which changes to the physical brain, such as drugs, head wounds, and strokes, as well as diseases such as epilepsy and Alzheimer's can have an impact on how the mind works and on a person's feelings, thoughts, and the sense of identity (Greenfield, 2000; Pfaff et al, 2022). Furthermore, researchers have increasingly acknowledged that the brain does not work in isolation. As the neuroscientist Susan Greenfield (2000:174) stresses, the brain is 'in constant two-way traffic with the rest of the body'. For her, the mind is the personalized brain, but unlike Damasio (2004), she believes the mind to be synonymous with the self: 'After all, if *mind* is the personalization of the brain, then what more, or what less, could *Self* actually be?' (Greenfield, 2000:185–186). This personalization takes place through the course of life, which turns a generic brain into a personalized mind, a process through which individual experiences and the meanings of each conscious moment are associated with the formation of networks of cells in the brain (Greenfield, 2000:164). However, the brain does not start life as a tabula rasa, as it knows from the beginning how to manage the organism, that is, 'how the life process should be run and how a variety of events in the external environment should be handled' (Damasio, 2004:205).

The relationship between materiality and subjectivity can also be traced to the notion of plasticity. Plasticity is a key feature of the brain, as its neural networks are always changing and reorganizing, with the implication that the constitution of the self, and our sense of it, are in continuous flux. William James is credited with first using the term to indicate the association between changes in nervous paths with the establishment of habits (Berlucchi and Buchtel, 2009). It is commonly believed that the child's brain goes through a major process of change and development, but research has shown that this process continues throughout our life (BBC, 2023).

The notion of neuroplasticity suggests that the brain is being continually rewired, and with it the human subjectivity is always a process of change and transformation via its interaction with the environment. This is a concept that combines materiality and transformation, allowing a dialectical understanding of the self as both embodied and altering (Malabou, 2010). It is not merely about the changes in form, as indicated in the term plastic arts, but also about the alteration of the contents and character, which in the brain takes place through continual rewiring. It is not merely about accidental and emergent changes, but also about purposeful changes that could lead to changes in form and content. As researchers have argued, the plasticity of the brain would allow certain purposeful exercises, such as learning a new skill or mindfulness, to be deployed for rewiring of the brain, indicating a level of agency at work (BBC, 2023).

Cognitive neuroscientists and philosophers of mind are criticized for confusing concepts and misapplying them to cognitive research. According to Bennett and Hacker (2007a), neuroscience investigates the empirical questions about the nervous system, and cognitive neuroscience sets out to explain the neural conditions that are necessary for perceptual, cognitive, cogitative, affective, and volitional functions. As such, cognitive neuroscience cannot engage with conceptual questions about the mind, memory and thought, which are the realm of philosophy and not amenable to scientific theorizing and empirical research. Neurological explanations cannot replace the wide range of psychological explanations about human activities (Bennett and Hacker, 2007a:7). Bennett and Hacker argue that this leads to ascribing psychological attributes to the brain, which is no more than mind-body dualism. They (2007b:20) accuse the first generations of neuroscientists to have fallen into the trap of the Cartesian dualism of mind and body, while the later generations, who claim to avoid this dualism, ascribe psychological functions to the brain, which is 'a degenerate form of Cartesianism'. As they argue, it is the human being as 'a psychological unity' who perceives, thinks, and feels, rather than the brain functioning as an independent and detached element: psychological powers 'are *attributes of human beings*, not of their parts – in particular, *not of their brains*' (Bennett and Hacker, 2007a:6).

What this debate shows is that neuroscience and cognitive science can shed light on the elements, structures, and functions of the brain from a third-person perspective, but they cannot engage in concepts and psychological attributes, or provide a first-person account about how a person perceives, feels, and thinks, and about the qualitative sense of self that is associated with these processes. The questions about the relationship between the brain and mind, and between the third-person scientific approach and the first-person experiential approach, have continued to remain as open questions (Robinson, 2007). The separation between mind and body continues to shape the structures of health provision, as reflected in the way physical health and mental health are kept apart. However, the debates about the relationship between mind, brain, and body show that while these three elements have been conceptually and experientially separated from each other, in practice they are fully integrated in a unified organism, and an analysis of the self cannot afford to consider them separately.

The knowledge that the self is produced through the agglomeration and transformation of neural networks and the interaction between brain, body and the surrounding world is important but not sufficient for our investigation of an authentic self. This biological account of the self shows how the transformation and agglomeration of neural networks in an organism lead to the processes of the mind and the emergence of consciousness as a centre of attention that is aware of itself and the world around it. These processes are the material events through which the self is formed and its physical relations with the body and the world around it are shown, but they do not engage with the question of what an authentic self might be. The search for an origin of the self would inevitably take us to the biological body and its interactions with the world. This sense of the self is continually changing, alongside the plasticity of the brain, the changes in the body, and in the surrounding environment. But this search cannot end there, as it does not engage with human subjectivity, unable to explain how this process can lead to an authentic or inauthentic outcome. In our journey, it is to the realm of psychology that we would need to turn next, searching for an understanding of that qualitative sense of the inner self.

## The fragmented self

In Freudian analysis, what is commonly known as self is divided into conflicting parts, each pulling it in a different direction. In his *Civilization and Its Discontents*, Freud (1991:253) first appears to equate the self with the ego: 'Normally, there is nothing of which we are more certain than the feeling of our self, of our own ego. This ego appears to us as something autonomous and unitary', with apparently 'clear and sharp lines of demarcation' on the outside. But Freud goes on to question this certainty and its sharp boundaries by looking at what lies beneath the surface. Freud (1991:253) suggests that the discovery made by psychoanalytic research has demonstrated,

> That such an appearance is deceptive, and that on the contrary the ego is continued inwards, without any sharp delimitation, into an unconscious mental activity which we designate as id and for which it serves as a kind of façade.

This division of the psyche into conscious and unconscious parts, as Freud (1962:13) explains in his book *The Ego and the Id*, is 'the fundamental premiss of psycho-analysis'. This division is based on a dualistic theory of conflict between repression and resistance: 'we obtain our concept of the unconscious from the theory of repression' (1962:15). The mind is described through a topographical hypothesis, split into two parts: the conscious and the unconscious, one repressing and the other being repressed (Freud, 2008). Freud uses a spatial metaphor to depict the mind, envisaging the conscious and unconscious parts of the mind as two adjacent rooms separated by a gate, arguing that 'these crude hypotheses, the two chambers, the door-keeper on the threshold between the two, and consciousness as a spectator at the end of the second room, must indicate an extensive approximation to the actual reality' (Freud, quoted in Bouveresse, 1995:24).

This spatial model holds consciousness to be 'the *surface* of the mental apparatus', both functionally and anatomically (Freud, 1962:19). The conscious ego is 'first and foremost a bodily ego' (Freud, 1962:26–27). For Freud (1962:25), the 'ego represents what may be called reason and common sense, in contrast to the id, which contains the passions', and he admits that this runs parallel with the familiar, popular distinctions. The term id is rooted in Nietzsche, who used it to refer to those parts of the humans that are impersonal and subject to natural law (Freud, 1962, editor's introduction). Aristotle (1991) defined humans as speaking animals, distinguishing the speaking part from the animal part, a separation that now appears to be reflected respectively in ego and id.

In Freud's model, the two parts of the psyche are in continuous conflict, whereby the repressed unconscious always attempts to find expression, as in dreams and neuroses, while the repressing consciousness always resists. Consciousness is attached to the ego, which is 'a coherent organization of mental processes' (Freud, 1962:17), exerting powerful control over some contents of the mind, excluding them from consciousness and repressing them into the unconscious. The task of psychoanalysis becomes helping to remove this resistance and bringing the repressed ideas into consciousness by representing them in words, thereby turning internal thought processes into perceptions (Freud, 1962:23). In other words, psychoanalysis sees its task as healing the patients through self-knowledge, made possible through shedding light on the dark corners of the mind, which are brought into consciousness by turning them into verbalized perceptions.

In the evolution of his theories, Freud comes to the view that the unconscious is not entirely made of repressed elements, and there is even an unconscious element in the ego. Attempting to explain the unconscious sense of guilt and higher value mental functions, he proposes the concept of 'ego ideal', or 'super-ego'. At the deepest level, mental processes are rooted in the id, from which first the ego, and then the superego, emerge. The ego is 'a specially differentiated part of the id', as 'the representative of the external world to the id' (Freud, 1962:38). The ego ideal is a part of the ego that is 'the representative of our relation to our parents', and therefore stands in contrast to the ego by representing the internal world, the id (Freud, 1962:36). The superego is the internalized voice of the parents, teachers, religion, and others in authority. It is part of the unconscious mind, and its tensions with the performance of the ego generate a sense of guilt. The conflict between the ego and the ego ideal is a reflection of 'the contrast between what is real and what is psychical, between the external world and the internal world' (Freud, 1962:36).

As Freud (1962:36) put it,

> Through the forming of the ideal, what biology and the vicissitudes of the human species have created in the id and left behind in it is taken over by the ego and re-experienced in relation to itself as an individual.

In doing so, 'What has belonged to the lowest part of the mental life of each of us is changed, through the formation of the ideal, into what is highest in the human mind by our scale of values' (Freud, 1962:36).

The result is a three-part theory of mind, made of id, ego, and superego, in which biological desires and social norms are both located in the unconscious, pulling the ego apart in opposite directions. The id is 'totally non-moral', the ego 'strives to be moral', and the superego 'can be super-moral and then become as cruel as only the id can be' (Freud, 1962:54). The superego appears 'essentially as a sense of guilt (or rather, as criticism)' (Freud, 1962:53), and the ego 'carries out repressions in the service and at the behest of its super-ego' (Freud, 1962:52). But ultimately the ego is 'a poor creature owing service to three masters and consequently menaced by three dangers: from the external world, from the libido of the id, and from the severity of the super-ego' (Freud, 1962:56).

The ego as the sense of the self is always in conflict with these unconscious forces that constitute it and are constituted by it. The ego exerts its agency by resisting and repressing the desires generated by the id, but still, it falls short by the demands of the superego, which keeps watch over the conduct of the ego in its dealings with the id and the world. As the external face of the mental processes, the ego may appear to be stable, but this façade is hiding a conflicting, tense interior, a battleground between instincts and norms, between desires and conscience. Despite evolving into a tripartite structure, the Freudian model of the mind remains dualistic at its heart, as the title of the book *The Ego and the Id* suggests, based on a conflict between the internal and external, biology and society, repression and resistance. This Freudian model undermines the Cartesian sense of the self as the only measure of certainty, as well as the Kantian notion of personal autonomy and other-worldly origins of the categorical imperative. Instead, the self is fragmented and uncertain, while the categorical imperative comes from the superego, which internalizes the society and its norms (Freud, 1962:35). This unstable, conflictual, and divided self would hardly provide a clearly identifiable reference point for a sense of authenticity.

Despite his disagreements with Freud, Carl Jung also divides the mind into the conscious and unconscious parts, referring to the unconscious as a shadow that resides inside the mind. The unconscious is 'a second psychic authority besides the ego' (Jung, 2014:61), but it is 'by definition … of uncertain extent and constitution' (Jung, 2014:64), 'manifesting itself mainly in the form of contrary feelings, fantasies, emotions, impulses and dreams' (Jung, 2014:61). In explaining the title of his book, *Undiscovered Self*, Jung (2014:xii) admits that the title was selected by his American publisher, and he thought that the self is 'not really undiscovered, it is merely ignored or misunderstood'. Jung argues that people tend to confuse the knowledge of self with the knowledge of conscious ego personalities: 'Anyone who has any ego-consciousness at all takes it for granted that he knows himself. But the ego knows only its own contents, not the unconscious and its contents' (Jung, 2014:3–4). Therefore, what we think of as self-knowledge is limited and largely dependent on social factors, which are influenced by mass society and state control. Jung is also sceptical about the suitability of just relying on scientific approaches for self-knowledge, as they are based on universal theories and statistical methods, whereas self-knowledge requires getting to know individual facts: 'There is and can be no self-knowledge based on theoretical assumptions, for the object of self-knowledge is an individual – a relative exception and an irregular phenomenon' (Jung, 2014:5). Rather than being universal and regular, for him what characterizes the individual is its uniqueness.

The individual is the 'carrier of ... consciousness' and 'the only immediate manifestation of the psyche' (Jung, 2014:34). But 'our conscious activity is rooted in instinct' (Jung, 2014:49), and therefore the individual 'remains a mystery to himself' (Jung, 2014:32), fearful of what may be discovered in the unconscious. The 'world of unconscious instinct is dominated by sexuality and the power drive (or self assertion)', and so Jung identifies two fundamental instincts: preservation of species and self-preservation (Jung, 2014:54–55). According to Jung, the 'ultimate foundations of the psyche' are represented by the 'archetypal forms', which are 'grounded on the instincts and giving expression to them' (Jung, 2014:35). These archetypal forms cannot be grasped intellectually or erased, as destroying one of their manifestations would be followed by their reappearance in an altered form. Instincts are original and hereditary, and their form is similarly age-old and archetypal, which is 'even older and more conservative than the body's form' (Jung, 2014:49). Instinct is not 'a blind and indefinite impulse' but 'attuned and adapted to a definite external situation', which gives the instinct a 'specific and irreducible form' (Jung, 2014:49). These archetypal forms are 'primordial patterns of ideas that were given us a priori', and their constant adaptation to the environment and its changes constitutes human knowledge (Jung, 2014:49).

Jung argues that in modern society, the dominance of scientific rationalism has created an abstract concept of the human and the world, and what is considered as self-knowledge is in danger of losing the inner shadow and identifying the self with a 'fictive personality' (Jung, 2014:59). Without acknowledging this instinctual nature, the modern human is plunged into a 'conflict between conscious and unconscious, spirit and nature, knowledge and faith', with possible pathological consequences (Jung, 2014:58). For Jung, knowing the self would mean going beyond the conscious ego and acknowledging the instinctual nature of the human. This would require getting to know the archetypes and instincts, but if these are not within the grasp of consciousness, knowing the self would always remain incomplete, acknowledging the existence of an unknowable shadow. Rather than making idealistic demands, therefore, the recognition of this shadow would lead to a sense of modesty and the acknowledgement of human imperfection, and how it is 'weak, helpless and in need of support' (Jung, 2014:73).

The Jungian model of the self, therefore, is a dualistic model similar to Freud's, based on a distinction between the conscious and unconscious. According to these psychoanalytic theories, a part of the self remains always beyond reach, and can only be understood through its indirect manifestations. For Freud, knowing the self would mean going beyond the ego and getting to know the id and the superego, a repressed domain of instincts, desires, and rules hidden in dark corners and shaping much of human thinking and behaviour. But here too, the id and the superego cannot be brought to consciousness. The self remains only partially known, without reaching its deep but important and powerful recesses. The implications of these models of partitioned and fractured mind are considerable for our investigation into authenticity, whereby the core of a self is always fragmented, hidden, and unreachable.

As we saw in the last section, cognitive neuroscientists have been criticized for confusing biological investigations with philosophical concepts (Bennett and Hacker, 2007a). The same critique has been raised against psychoanalysis. Despite his fascination with the charms of psychoanalytic theories, Ludwig Wittgenstein doubted their status as a science, as Freud and his followers had claimed them to be. As Wittgenstein told a friend, 'Freud is constantly claiming to be scientific. But what he gives is *speculation,* something prior even to the formation of a hypothesis' (Quoted in Bouveresse, 1995:7). Wittgenstein considered the unconscious to be a hypothesis and an unsatisfactory part of Freud's theory (Bouveresse, 1995:26); it was a myth and a manner of speaking rather than a scientific discovery (p.xvii). As Wittgenstein wrote, 'What Freud says about the subconscious sounds like science, but in fact it is just a *means of representation.* New regions of the soul have not been discovered, as his writings suggest' (quoted in Bouveresse, 1995:27). The methods of psychoanalysis, Wittgenstein argues, are not experimental but aesthetic: 'As in aesthetics, things are placed side by side so as to exhibit certain features' (Wittgenstein, quoted in Bouveresse, 1995:27). If psychoanalysis interprets a dream as the disguised fulfilment of a wish, it is not a scientific exposition in the conventional sense of the word, but engaging in an act of interpretation and persuasion; it is the extension of a concept to a theory, which is closer to philosophy (Bouveresse, 1995:xvii, 124).

If the Freudian id is assumed to be the origin of the self, it is deeply hidden in the unconscious, only manifested indirectly; thereby it remains beyond reach, not offering itself as a platform for any concept of authenticity. At the other end of the spectrum, the superego is a set of general norms and conventions, devoid of specificity and authenticity. The ego is torn between these two ends, never able to overcome this fragmentation and construct a self that is whole and free from these pressures. The self is shown to be a battleground, where the warring parties can never reconcile to form a unity, an origin that could serve as a solid core and a reference point for authenticity.

## The imagined self

Jacques Lacan brings structural linguistics, which were highly influential in the mid-twentieth century, into Freudian psychoanalysis. The central line that connected Lacan's various research projects was the idea that a 'human being is implicated in structures', as shown in the relationship between the human subject and language, whereby the human subject is made to be 'passive, mastered, reduced to being a "serf of language"' (Burgoyne, 2003:70, 73). Rather than a dark corner full of libidinal impulses, Lacan saw the unconscious to be structured like a language, that is, the 'natural languages' that are 'effectively spoken by human groups' (Lacan, 2006:371). Rather than moving closer to medical sciences, his approach to psychoanalysis was revitalizing it through epistemology (Rabaté, 2003:xiii). Lacan's central insight was that 'the body and the "soul" ... were connected not via Descartes' pineal gland but simply by language' (Rabaté, 2003:xii). Rather than an instrument that we can use as we wish or a tool for self-expression and communication, language is 'fundamentally inscribed in the real' (Soler, 2003:91). The relationship between the body and language is the mystery that he sets out to explore. As Lacan puts it, 'The real, I will say, is the mystery of the speaking body, the mystery of the unconscious' (quoted in Soler, 2003:87).

This idea was pursued in 'the relentless exploration of a single field, that of the speaking id', which was about 'the interaction between the suffering body in its manifold symptoms and the suffering mind when it stumbles in parapraxes and unconscious delusions' (Rabaté, 2003:xii). The speaking id comes close to the Aristotelian definition of humans as speaking animals, via a Hegelian notion of social mediation and recognition: 'what we must say, following Aristotle's lead, is that it is not the soul that speaks but man who speaks with his soul, on the condition that we add that he receives the language he speaks and that, in order to bear it, he sinks more into it than his soul: he sinks into it his very instincts...' (Lacan, 2006:392–393). This formulation also comes close to Heidegger, whose work Lacan had translated into French, and who in his 1947 Letter to Humanism, had declared language as the 'house of Being' (Heidegger, 2011:161).

In Lacan's work, the Freudian distinction between the unconscious and conscious mind is linked with the Saussurian distinction between signifier and signified. The linguistic insight was 'to see the signifier as the determinant of the signified', and psychoanalysis reveals the bar that separates the signifier from the signified, the unconscious from consciousness, showing the structure of the subject to be 'a discontinuity in the real' (Lacan, 2006:678). The subject is not a unified, self-conscious, or integrated personality functioning at the centre of human thought, but an unconscious that 'inhabits the mind as an elusive agency, controlling yet uncontrollable' (Nobus, 2003:61). The unconscious is the 'discourse of the Other' and therefore 'as a systemic social formation, a hoard of words, names, and sentences out of which collective utterances are made', which also accounted for 'my own singularity, thanks to the agency of the specific condensation of signifiers that appears as a symptom, that is, *my* symptom' (Rabaté, 2003:xii). Lacan uses the terms 'the big Other', and the 'symbolic order' for this linguistic structure of the mind: it is 'society's unwritten constitution, ... the second nature of every speaking being' (Zizek, 2006:8).

The bar between the signifier and signified creates an obstacle, a lack, which becomes a central concept in Lacan. For Lacan, this is a primordial condition, indicating that human subjects are always in a state of incompleteness, a lack, which triggers desires that can never be satisfied. The lack is rooted in children's early experiences of separation from their mother, entering the social/linguistic world, and learning about their identity through their own image in a mirror and in the eyes of others. Humans are always searching for completion, which they can never reach (Luepnitz, 2003). The self and the knowledge of the self, as experienced in the child's mirror stage and in the acquisition of language, is always experienced as an unbridgeable gap. At an early age, at a time of complete dependency and weakness, human subjectivity is made by others, by their language, desires, and jouissance. The subject becomes formed as a divided entity, driven by these forces that are internalized in the form of the unconscious.

By introducing linguistics to psychoanalysis, Lacan turns the Freudian triad of id-ego-superego into a triad of real-imaginary-symbolic. Where the superego represents the social norms and authority, the symbolic order represents the linguistic structures and the social world. Later Lacan calls the symbolic order the 'big Other', the point from which I judge myself (Zizek, 2006:80). Where the id represents the sexual and death instincts, the real indicates the materiality of being and the unreachable depths of the psyche. It is an elusive concept that Lacan changes over time and marks the limits of the imaginary and the symbolic (Johnston, 2024). Where the ego is the consciousness torn between the biological and social forces, the imaginary is a fiction that is constructed in the void that separates the symbolic order from the real. The imaginary is what Lacan calls the 'small other', the idealized image of the self, the ideal ego (Zizek, 2006:80). Where the knowledge of the ego was limited and insufficient, as it could not reach the unconscious, which stood in the shadows, knowledge of the imaginary would be the only knowledge that we have of the self, which nevertheless remains limited and unreliable.

Lacan argues that the subject is 'decentred', as the subjects' fantasies and desires are not really theirs, but a reflection of what they think is expected of them by others (Figure 2.1). 'The original question of desire is not directly "What do I want?", but "What do *others* want from me?"' (Zizek, 2006:49). As a result, the subject becomes 'empty', and 'divided, as the agency that is not able to assume the very core of his or her inner experience' (Zizek, 2006:54).

From the psychoanalytic perspective, therefore, a complete and recognizable self does not exist or is not within reach. What is usually considered as the self is the ego or the imaginary register, which is a mere mirage. Knowing the self requires the hard work of going beyond the surface, and looking for what lies behind, in the social norms and symbolic orders in which a person has been raised and lives, in the biological drives and forces of life that are continually at work but are repressed, in the voids that exist between the biological and the social worlds. Knowing this complex topography of the unconscious mind, however, can never be complete, as it is not a fixed and

*Figure 2.1* Do I know what I want, or is it what others want me to desire? (New York, United States)

stable entity, or one that could be stabilized in the light of consciousness. Our search for an authentic sense of the self, therefore, would only take us to an unknown and unknowable world of the unconscious, with its bundles of unending desires, recurring archetypes, and unbridgeable voids. As soon as we recognize something as a sign of our inner self, it can shift its shape and re-emerge as something else. The search for the self as a solid ground and a basis for authenticity takes us to unending journeys in a labyrinthine unconscious world. Knowing the self would end up rediscovering the social norms as they are played out in a single biography, without being able to see the deep recesses of the mind, bewildered in the gaps between the body and the linguistically structured mind.

Freudian ideas are seen by some as the third major step in decentring humans from their age-old, high pedestal, which had secured their position as the centre of the world. The Copernican step denied the centrality of humans in the universe, shattering the ancient model of cosmology and showing the earth to be a mere planet revolving around the sun. The Darwinian step denied the exclusive place of human in the world, rejecting their separation from other animals, even if still maintaining the location of the human at the top of an evolutionary hierarchy. The Freudian step finally decentred the conscious human mind, arguing it to be the mere surface of a hidden, stronger world of unconsciousness that lay underneath and ruled over it (Descombes, 1995).

Questioning the unity of the self is not new. Various theories throughout ages have analysed the mind into its constituent parts, as exemplified in Plato's tripartite theory of reason-emotion-passion (Plato, 1997) or Kant's tripartite cognitive powers of understanding-judgement-reason (Kant, 1987:38), among others. Many of these schemes have been based on the presupposition of a single centre that would combine these mental faculties into a single self. There are those, however, who have questioned the existence of such a unified centre.

In the sceptic, empiricist tradition, the self is no more than a bundle of shifting perceptions. In the early eighteenth century, the Scottish philosopher David Hume (*A Treatise of Human Nature*, I, iv, 6) wrote of the fragmentary nature of the sense of self and personal identity. According to Hume, the sense of self is always mediated through perceptions: 'when I enter most intimately into what I call *myself* I always stumble on some particular perception or other, of heat or cold, light or shade, love or hatred, pain or pleasure' (Hume, 1826:321). As a result, 'I never can catch *myself* any time without a perception, and never can observe any thing but the perception' (Hume, 1826:321). And these perceptions are multiple and always changing: 'But self or person is not any one-impression, but that to which our several impressions and ideas are supposed to have a reference' (Hume, 1826:321). Impressions are not constant and unchanging, and we have many impressions which follow each other and are never present at the same time; they are 'in a perpetual flux and movement'. Hume concludes that the self is nowhere to be found: 'It cannot therefore be from any of these impressions, or from any other, that the idea of self is derived; and consequently there is no such idea' (Hume, 1826:321).

In the much older Buddhist tradition, the self does not exist, and what we know as the self is no more than a bundle of energies. The Buddha questioned the existence of the self, replacing the self-knowing soul with an abstract thirst for life. Life is made of five aggregates that are a combination of physical and mental energies. There is 'no permanent Self (*Atman*), individuality, or anything that can in reality be called I' (Rahula, 1974:26). Instead, there is 'the will to live, to exist, to re-exist, to continue, to become more and more' (p.31). As the ultimate aim of this philosophy is to stop the cycle of returning to life, the will to exist must be overcome by a stronger will to desist. 'As long as there is this "thirst" to be and to become, the cycle of continuity (*samsara*) goes on. It can stop only when its driving force, this 'thirst' is cut off through wisdom' (Rahula, 1974:34). Rather than a bundle of perceptions, which tends to be united in the centre of attention, the non-self here is a bundle of energies and a drive for continuity.

What does this picture say about the urban society and what are the lessons for urbanism? The urban society lives at the intersection of social rules and individual behaviour. While individuals navigate the conflicts between rules and desires, between social pressures and individual urges, the psychoanalytic perspective recommends that the inbuilt drives and desires should be acknowledged and verbalized, rather than repressed. This generates a pressure for openness and revelation, which can take many forms, from healing the neurosis to praising the freedom of expression and to licensing the free flow of never-ending, never-satisfying desires. In response to the latter concern, Jung argues that the fear of anarchic instincts being let loose is an exaggeration, referring to the existing mental and social safeguards: 'Above all, there is the natural cowardice of most men to be reckoned with, not to mention morality, good taste and – last but not least – the penal code' (Jung, 2014:38–39).

The psychoanalytic investigation shows that the self is not a single and coherent unity, but a dualistic structure made of conflict, and built on deep but unknown and unknowable foundations. However, the attempts of the id at resisting against the repressions of the ego, urged by the external pressures of the superego and the symbolic order, may give the impression that the authentic self is an entirely internal phenomenon. Any external force, such as the social habits and norms, or any common infrastructure such as the language, may be thought to be alien to this authentic self. The dualism that is implied in this research places the social element as the repressive force, and the biological element as the repressed, thereby in need of acknowledgement and assistance. It seems to imply that authenticity resides in the id, which is under pressure from the inauthentic superego and the symbolic order.

At least this seems to be the lesson that the popular media and self-help gurus offer from the psychoanalytic research. Whatever is inside you is authentic and your true you, but anything from the outside is merely an imposition, something that has no affiliation with your true self. However, as we have seen, the psychanalytic tradition also draws attention to the significance of the social elements in the constitution of the psyche. It shows the complexity of human subjectivity, and how it is constructed from birth under the influence of its social environment. For finding the meaning of authentic self, therefore, we need to investigate that social element, which is taken up by the fields of social psychology and social philosophy.

If authenticity means being true to yourself, and if this self is shown by sceptic empiricists and psychoanalysts to be unstable and multi-layered at its core, how can you be true to something that is so ambiguous? If authenticity means referring to an origin, and this origin appears to be an unreachable place in an undiscoverable labyrinth or even a non-existing entity, how can authenticity be even possible? If being true to yourself means unifying the inside and outside, and this inside is a fragmented bundle of personal perceptions or socially constituted fantasies, or it is not open to the full light of conscious knowledge, how can such a union be achieved? Can it be found on the outside, in the social realm?

## The mediated self

The social character of the self is largely revealed through its presentation to others, a notion that has long been known. The famous phrase 'Know Thyself' is often attributed to Socrates. This was in fact an ancient inscription in the temple of Apollo at Delphi, which Socrates admiringly quotes in various dialogues (Plato, 1997). In one dialogue (Plato, *Philebus*, 48d–49c), Socrates analyses the meaning of this inscription through its direct opposite, that is, not knowing oneself, which is manifested in three ways: in relation to money, physical advantage, and virtue. When people boast that they are richer, taller and handsomer, and wiser than they actually are, they show that indeed they do not know themselves and have 'a false opinion about themselves'. When this delusion is combined with weakness, others laugh at them, but when combined with power, they become dangerous. This interpretation shows that for Socrates, self-knowledge is not merely an act of private reflection into the psychological structures of the mind or excavating its deep recesses, but it is closely related to a person's place in society. It is not merely about knowing your mind, but also your body, belongings, and social position. It is about the relationship between self and other, between individual and society.

In another dialogue, with disputed attribution to Plato (Plato, *Alcibiades*, 133c), Socrates equates the self with the soul, as distinctive from the body, and knowing oneself becomes the knowledge of the soul and 'being self-controlled'. He advises Alcibiades that 'what you need to get for yourself and for the city isn't political power, nor the authority to do what you like; what you need is justice and self-control' (Plato, *Alcibiades*, 134d). These two discussions of self-knowledge show that for Socrates, the mind is not detached from the body or social context. Knowing yourself becomes knowing your abilities and limitations: understanding your physical, social, and moral characteristics accurately, presenting them truthfully, and cultivating yourself to avoid excess. Socrates contrasts knowledge with ignorance, and he privileges knowledge over pleasure, reflecting the Platonic worldview in which an intelligible realm of ideas stands high above the material and sensual world. What is necessary for the good life is to become aware of that intelligible world, and to ensure that it rules over the sensible world, keeping it under control.

These two Socratic interpretations of self-knowledge look at the relationship between self and other. They are both recommendations for self-control, limiting the yearning for over-representing the characteristics of the self and letting its desires loose. They articulate this relationship from two angles. The first interpretation urges the union of the inside and the outside in the public presentation of a self. The social presentation of the self, therefore, is expected to reflect the actual circumstances and characteristics of the individual. It is an exercise in honesty, which equates public performance with personal qualities. The second interpretation invites social considerations such as justice to rule over individual desires. Both interpretations show an ethical framework in which respect for others, through sincerity, moderation, and self-control, is placed well above the desire for power, self-gratification, and excess.

In requiring a reduction in pretension and excessive playacting in social encounters, this appeal to being true to yourself has always been one of the ethical principles of social life. According to this principle, which Socrates explains well and is also reflected in almost all of the world's moral and ethical schemes, society condemns lying and duplicitous behaviour, even if almost everyone may sometimes lie, exaggerate their qualities, or perform in ways that are completely at odds with their own view of themselves. This has long been understood as the ethical demand for sincerity, a demand that in the modern period was metamorphosed into a concern for authenticity (Trilling, 1972). When the Socratic perspective is confronted by the psychoanalytic perspective, it may be interpreted as the triumph of the superego, where the symbolic order prevails, and Stoic repression of desires would run the risk of causing neurosis.

The primacy of the social environment in the construction of the self, which was taken as an ethical imperative for centuries, became questioned by the emergence of the idea of the individual in modern society, which evolved through a long process that had started in the Middle Ages (Coleman, 1996; Taylor, 1989). Renaissance humanism and enlightenment rationalism raised the status of the individual to a level of omnipotence that could replace the authority of the church. According to this view, therefore, the solid base on which a sense of authenticity could be built would lie in individuals and their ability to think for themselves. As it was soon discovered, however, it is impossible to envisage a person in a void, detached from any context. A prominent critique of purely individual subjectivity is developed by Hegel. According to Adorno (1984:167), Hegel proposes the dictum that 'there is nothing between heaven and earth that is not mediated'. Rejecting the Cartesian dualistic thinking, and following Spinoza's monism, Hegel advocated a dialectical process that linked the subject with the world (Siep, 2014). For him, consciousness is rooted in a social context, as played out in the relations of intersubjectivity and recognition (Siep, 2014:90–92).

With his emphasis on temporality, Heidegger historicizes human subjectivity, banishing the centrality of consciousness and the concept of selfhood (Gadamer, 1976:50). According to Gadamer, 'all understanding is self-understanding' but not by a self in control of itself: 'The self that we are does not possess itself; one could say that it "happens."' (Gadamer, 1976:55). For him, self-understanding is an ongoing dialogue with oneself, anticipating communication with others (Gadamer, 1976). For Lacan, meanwhile, communication with others is not with particular individuals, but with the symbolic order, which is the anonymized other (Zizek, 2006). These approaches, and many others, which all draw on the Hegelian notion of social mediation and recognition, indicate the significance of the other in self-understanding,

*Figure 2.2* Looking for the self as a detached entity would inevitably lead to others (Bologna, Italy)

that the sense of the self is a never-ending, socially mediated process. Therefore, as the search for the true self, authenticity would always reflect the expectations that the other, personal or impersonal, would have from us. The origin or point of reference that would shape the sense of authenticity would not be just excavated from deep inside the mind and body, but it would largely be found somewhere else on the outside. The relationship between the inside and the outside shapes self-understanding and the search for authenticity (Figure 2.2).

There are many circumstances in which some material or cultural features of a person or group are so suppressed by the social environment, that they are always forced to hide those features, such as gender, language, or beliefs. They may always live a double life, an inside that they feel is their true self, and an outside in which they perform to be someone else, denying their sexuality or cultural beliefs. The desire to be true to yourself would mean dropping the forced performance and being able to express themselves more freely. This is a major tension in the lives of minority populations who live among oppressive majorities, who either deny the existence of such minorities or prevent them from acting according to their own biological or cultural needs and wishes. If the outside is inauthentic, it is due to the social pressures for conformity and the denial of diversity as an existential condition of any society. If authenticity demands the unity of the outside and inside, it could easily mean creating the conditions in which the diversity of the population can find open expression, rather than being suppressed. Heidegger called language the house of being, so the social groups who are not allowed to speak in their mother tongue in public are deprived of accessing a key part of their life. The original populations, in the lands around the world that were colonized by the European settlers, were forced to change their names, language, beliefs, and behaviour, and were pushed out of their ancestral lands, becoming alienated from the forms of life that hitherto they had lived.

As Agamben shows, the ancient Greeks used two terms to refer to life: '*zoē*, which expressed the simple fact of living common to all living beings (animals, men, or gods), and *bios*, which indicated the form or way of living proper to an individual or a group' (Agamben, 1998:9). Agamben is interested, following Foucault and Benjamin, in the encounter between the bare life of *zoē* and modern politics (p.10). This dichotomy runs parallel with the relationship between the embodied self and the social self, between biology and social mediation. This encounter is one of the key concerns of the discourses of authenticity, both visible within European societies and in their encounters with the world that they tried to conquer. The quest for authenticity appears to be a search for the *zoē* as a purer form of life, sometimes praised in what seemed to them as simpler societies. But at the same time, whenever these so-called purer forms of life were found, they were forced to change, becoming assimilated to the forms that the colonial powers imposed on them. This was never a project of praising the purity of life, but imposing one form of life on another, one *bios* on another, by design or by accident.

## The reflexive self

The self has been a major theme in social psychology (Baumeister, 1999; Branaman, 2001; Gilovich et al, 2016; Hogg and Vaughn, 2018; Rousseau, 2002; Smith et al, 2015). The concept of the self is described as 'the set of all an individual's beliefs about his or her personal qualities' (Smith et al, 2015:96). The ability for reflexive thought has been considered to be a distinctive human ability, which is 'the ability to think about ourselves thinking'; it means 'that we can think about ourselves, about who we are, how we would like to be and how we would like others to see us' (Hogg and Vaugh, 2018:118). The idea of self and identity is an integral part of this reflexive thinking; they are 'cognitive constructs' that are influenced by society and in turn influence our social interactions and perceptions (Hogg and Vaugh, 2018:118).

The idea of the self as a cognitive construct is part of a wider theme in social psychology that sees reality as subjective and socially constructed through interaction with the family, group, situation, and culture (Gilovich et al, 2016:68). The social character of the self has led to longstanding debates in social psychology as to whether the self is an individual or a collective construct. As the self emerges in interaction with the social environment, the self itself is seen by some as a collective self, difficult to distinguish from others in the same group.

Reflexivity is an integral part of human subjectivity: we can look at ourselves as if we are an outside observer. This observation is not merely an act of understanding but also judgement and action. The ability to keep a distance from our thoughts and behaviour, to know them and to evaluate them, is one of our primary methods of locating ourselves in the world. Reflexivity functions as a feedback loop, which allows me to relate to my environment, to learn about it, to accept some aspects of it, and to reject others, while also adjusting and changing my own thoughts and behaviour and trying to improve my place in the world. Reflexivity is an attempt to assume an entity that is outside me, to observe and evaluate that entity, an effort to detach my thoughts from myself and from the world around me. The ability to separate my thoughts from myself, to become conscious of my consciousness, also allows me to separate myself from my environment, thinking about it as a set of phenomena that is close-by but detached from me.

We can raise critical questions about reflexivity: that we cannot really become an outside observer of ourselves or even of our environment, as we are an integral part of both. This critique may show that what we see when we look at ourselves may be a mere surface, a projection of our thoughts, a reflection of social forces, or a momentary awareness that continually changes with our moods and circumstances. This line of questioning may show us that what we consider to be a good understanding of the self may be no more than an illusion or a narrative fiction. This is not to say that our sense-perception about the world around us is necessarily an illusion but to say that our awareness of our own thought and behaviour cannot be as detached or as accurate as we imagine them to be. The image of ourselves is a mental image that changes with our moods and in relation to our circumstances.

However, what is not an illusion is the capacity and the need for reflexivity, which is a tool that enables us to live in the social and material world. As a feedback mechanism, which develops through experience and observing the reactions of others to our behaviour, we develop the skills of living in the world. We look at ourselves through the eyes of others, or at least through the internalized impression that we have of these others, who may be particular persons of significance, and/or a more generalized sense of society. The sense of the self will be different from the presentation of the self to others, but both are interrelated processes that are shaped through living in a social world. In other words, reflexivity is a mechanism that makes living in society possible. It shows that personal identity is dialectical, constructed through interaction with the world around us.

The identity of the self emerges through the relations of similarity and difference as played out in society (Jenkins, 1996; Madanipour, 2013). Disparate fragments of memory are brought together to construct a narrative that can make sense of these past experiences and their traces in memory. It is an attempt to give coherence to what is primarily multifarious and fragmented. And it aims at presenting the narrator in a good light, to enhance the narrator's social standing and self-esteem. The sense of self is made of a core category and a series of attributes. The core category is a narrative that maintains a degree of coherence and stability for self-identity, while the attributes are more flexible, adopted, and adjusted according to the situation. The core narrative is itself made of the attributes that we accumulate throughout life, which are edited and turned into a more or less meaningful story about our sense of self.

Personal identity may be analysed as a narrative identity (Ricoeur, 1995). Being true to myself, therefore, would mean being true to the narrative that I have constructed about myself, either explicitly or implicitly. But this narrative is not entirely based on introspection. Introspection is an unreliable and insufficient source for self-knowledge, but memory plays an important part in its development, and therefore the self-concept is also constructed by seeing oneself from the outside, in the same way that a concept of other people's character is formed. This is done by observing other people's behaviours to learn their characteristics, by reflecting on the reactions by other people to one's own behaviour, and by comparing oneself to others (Smith et al, 2015:96).

What are the implications of this significance of the social environment for the construction of a self-concept? Does it mean that the search for a self would inevitably lead to the discovery of society, rather than anything unique or significant about the individual? Does it mean that any aspiration towards authenticity would evaporate as nonsensical? As in our journey inside the mind, are we facing shaky ground made of moving layers, rather than the solid base for developing a sense of the self, its identity, and its authenticity? It may be in the spaces in between these two uncertainties that a sense of authenticity might be found, where those narratives of self are constructed and presented, seeking recognition and developing self-esteem. This might be a challenge to, and an opportunity for, the idea of uniting the inside and outside. Rather than deep recesses of the mind, the sense of the self may lie elsewhere, on the borderline between the inside and the outside, where the self becomes observable in its manifestations. It is on this borderline that the inside and the outside can be seen together, and the sharp dividing line between the two realms might loosen, seeming to be no more than a convention.

## Inside and outside spaces

The demand to be true to yourself entails loyalty to an inner truth, but this inner truth is in fact largely a reflection of the outside world. This interdependence of the inside and outside shows authenticity to be dialectical and relational. The relationships between the self and others find spatial expressions such as the inside and outside of the human body, the personal and impersonal encounters, the indoor and outdoor spaces, and the private and public spaces, which are all parts of any discussion of the implications of authenticity for urban space and society. Does the quest for authenticity mean that the deep corners of the human mind, personal interests and desires, and private spaces should prevail in the city, making the outside world of others a mere subordinate? Does it mean that anything on the outside is inauthentic? Does it indicate the inside spaces are necessarily authentic?

If authenticity indicates being true to the self, and the self of an urban society is presumed to lie in an inner core in the households, would it mean that it is behind the closed doors of the private spaces that we should seek authenticity? But we have seen the limitations of seeking authenticity in detachment from the world, how the individual self, if it can be found, is not homogenous, unitary, and detached but split into conflicting parts, always in interaction with the surrounding environment. In a similar way, the private spaces of the home are not fixed, pure, and secluded, but spaces in which the members of the household are related to each other and to the outside world (Madanipour, 2003). The members of the household may drop some of their masks in the intimate space of the home. But like its individual members, the household as a group is also formed through social conventions and expectations, which are intensified through institutional arrangements

*Figure 2.3* Is every doorway separating an authentic inside space from an inauthentic outside? (Alexandria, Virginia, United States)

and technological instruments. The interior life of a house, especially in the context of the changing structure of the household, can show some similarities to the conflictual interior of the psyche, largely shaped through social conventions. The inside spaces could be as much exposed to the accusation of inauthenticity as the outside spaces are (Figure 2.3).

The social environment is performative and explicitly subject to social norms and conventions, and therefore may be seen as inauthentic. The everyday examples include politeness to strangers, and aesthetic awareness in the presence of others: the attempt to look good, wearing jewellery, makeup, and clean, presentable clothes, all as part of the attention to how a person appears in public, what the Italians call *Bella Figura*. In the urban environment, the examples include attention to the front facades, through good maintenance, placing flowerpots and well-kept front gardens, as well as setting up street decorations and embellishing the streets and squares. In extreme cases, differentiating the quality of the front and back facades may include keeping the old facades and changing the entire building in redevelopment projects (facadism). They are all examples of performance in the public environment, which, by being different from what goes on behind the scenes, could be considered exercises in inauthenticity. And yet, most of them are the everyday scenes of urban life, the building blocks of the urban environment.

The implications are that either we need to conclude that the public environment can never be authentic or to redefine authenticity as not being limited to merely prioritizing the inside. For example, there are authentic sculptures in the public environment, which are authenticated by reference to both the sculptor and the period and culture in which they were made. There are authentic streets in the urban environment, which are evidence of a particular period in history and a specific culture, as exemplified by the Roman streets that can still be found in many European and Middle Eastern cities. There are neighbourhoods in cities, in which a composition of activities, or a clustering of specific groups, gives that neighbourhood a distinctive character judged to be authentic. In these cases, the reference by which authenticity is measured is not an inner subjective or private space, but something beyond that in terms of space, time, and society (Figure 2.4).

The pandemic of 2020 first created a fear of the public environment, while also showing its central significance for social life and wellbeing. Before the pandemic, at a conference of cultural studies, the speaker gave some Lego pieces to the audience to construct their ideal public space. Some of these models were made in the image of a private space, which included most of the elements that you may find in a house's private, enclosed garden. This was the projection of the image of private space onto the public environment, turning it into a mirror of the private space. The private space may be associated with comfort and intimacy, and it may be where the search for authenticity leads to, but the transformation of the public environment into a private space would dismantle most of the platforms on which society is based.

*Figure 2.4* Would the public environment, with its range of people and objects, be simply dismissed as inauthentic? (Rome, Italy)

If the notion of authenticity requires a unification of the inside and outside, of the private and the public spheres, it may never materialize, as this unification is both undesirable and impossible. Life in the home can never be the same as life on the street, and patterns of human behaviour change according to the presence and absence of others, and their degrees of familiarity. This is true even in small villages, where everyone knows everyone else, and there is almost nothing to hide from one another. And yet, people keep boundaries between their personal and social lives, which enables them to live peacefully and comfortably. This is even more true in complex urban living, where the level of familiarity with others is reduced and the city is made of thousands and millions of strangers. In these cases, individuals adjust their behaviour to the crowded environment, which may include performing indifference towards others (Engels, 1993; Simmel, 1950), even if their livelihood is highly interdependent with these strangers. Rather than complete separation or unification of the two realms, acknowledging and negotiating their relations through a series of flexibilities and in-between spaces, which are semi-public and semi-private, are essential ingredients of managing life in the city.

The relationship between the inside and outside is a central concern in architecture and urbanism. The boundaries between public and private spaces regulate social relations; the character of these boundaries can be considered a good indicator for the character of the urban society. This is evident in cities where the boundary is hardened by high walls, barbed wire, and the threat of violence, or softened by the removal of walls and gates, appearing to be welcoming to strangers.

In much of the world and throughout history, a primary form of housing has been constructing spaces around a central courtyard, whereby the appearance of the house does not show its interior. This model of housing can be found from China to the Mediterranean Sea, from ancient Mesopotamia to today's Middle East, whereby the interior appears to be a different world invisible and inaccessible to the outside world. The façade or the entrance to the house may be simple or elaborate, but it is always different from the inside. Nowadays, this introversion has been widely transformed into extroversion, whereby much effort may be placed on embellishing the outside, which may or may not reflect the interior arrangements. These two gaps between the outside and the inside pretend the household to be more modest or more affluent than they really are. Judging by the standard of being true to yourself, most houses throughout history could be seen as expressions of inauthenticity.

In fact, both of these trends of interiority and exteriority mirror the social values of their society and era. If pretension to higher status is frowned upon, if this exposes you to unwanted risks, and if gender and class relations expect particular forms of behaviour, then the private space would be hidden from view. In contrast, the exterior of the house can show off the wealth of the household if explicit competition for status has become a safe option and an acceptable social norm. If we take authenticity to be a reflection of the society and epoch, as we shall discuss in later chapters, then both of these trends of disclosure or concealment of household material and social conditions should be considered authentic. In other words, the gap between the inside and the outside, between a concealed space and a disclosed one, maybe simultaneously interpreted as being authentic and inauthentic. The reason for this ambiguity is that, as in the case of self and other, the two sides are not expressions of two different worlds, but they show the different aspects of the same culture.

The rise of functionalism in modern urban society set out to reduce this gap between the interior and exterior. Modernism despised decoration and advocated transparency in the construction and parity between the outside and the inside. The building's elevation was to be shaped by its plan, not by unnecessary decoration, as it had been practised in the nineteenth-century European cities. When Paris and Vienna were transformed during the second half of the nineteenth century, attention to the embellishment of the facades was seen by the critics as an aesthetic attitude without functionality. The modernists argued that the logic of the interior should inform the appearance of the building. In this sense, modernism in architecture and urbanism would be a high point in advocating authenticity. It shows how paradoxically both the admirers and opponents of the modern industrial society, from modernists to Heidegger, could embrace authenticity as a central concern.

Modernism argued that the façade should reflect the plan, and the walls that separate the inside and the outside should be either removed or made transparent. Some works of modernist architecture create the appearance of a seamless flow between the inside and outside, where it seems that the space of the house has opened up to the outside, without any barrier, making the house an almost fully transparent space. A famous example is the Edith Farnsworth House designed by Ludwig Mies van der Rohe. It is a beautiful minimalist design that unites the human shelter and the open green spaces that surround it. However, it is important to notice that in such houses, which are located somewhere far from other buildings, privacy is maintained through distance. It is difficult to see these houses in dense urban areas, where many thousands of strangers pass through the streets, and where houses stand next to each other in crowded neighbourhoods. In these cases, the inside is protected from the intrusion of the outside, so as to keep the private life of the household safe.

The modernist unification of the inside and the outside was also the subject of experiments in socialist countries. The private space was reduced through the abolition of private property, and the intimate space of the house could be opened up to communal modes of living. In turn, the public space would be indistinguishable from the state space, as the state controlled much of social life. Public–private relations would be transformed through the subordination of both spheres to the state as the dominant institution in society. In the early years of the Soviet Union, many experiments in the various forms of art and literature could be seen as expressions of a new form of authenticity, which saw themselves at the service of a revolutionary social movement. However, with the rise of Stalinism, these experiments were suppressed in favour of the approved expressions of state ideology, where the notion of authenticity could hardly find space for development and expression.

Publicness and privacy are relational processes of setting, negotiating, and maintaining boundaries (Madanipour, 2003; 2023). So, it is possible to have a private conversation in the middle of the public street or to be exposed to public scrutiny while staying at home. The process of boundary setting, which separates the private and public realms through exposure and concealment (Nagel, 1998), appears to be at odds with the processes of authenticity, which advocate favouring the inside or uniting the outside and the inside. If authenticity is understood as a recourse to the self and to the private space, the public realm may be abolished or turned into a mere collection of private spaces. If authenticity is understood as the unification of the inside and outside, it comes into conflict with how society lives. However, as we found that the self and others are not sharply separated, these clear-cut distinctions and dichotomies do not necessarily occur. As the sense of the self is shaped through the relations between subjectivity and materiality, the relations between the inside and outside, the private and public spheres, are shaped interactively and dialectically, rather than categorical cases of rigid demarcation.

The widespread use of digital technology and social media may seem to soften or remove the sharp boundaries between public and private spaces, but they simultaneously undermine the possibility of authenticity (Figure 2.5). By creating an avatar, or by performing in front of the camera for uploading to the social media, a curated version of self is created for the others that is at once private and public. The private space is opened up to surveillance, commodification, and manipulation, thereby denying the possibility of privacy in any space. Simultaneously, the presentation of privacy in the public space of social media turns the public sphere into an endless array of performing private moments, rather than a place of intersubjective interaction, undermining the possibility of authenticity further. Rather than articulating the boundaries for a dialogue between the public and private spaces, between the inside and

*Figure 2.5* The boundaries between the inside and outside have been radically transformed through digital media (Beijing, China)

the outside, the blurring of the boundaries through digital media reflects the possibility of aggressive intrusion from either side.

### In search of a solid core

The common point of reference for the discourses on authenticity is the self. It is the basis for a wide range of activities in the popular culture that encourage people to be true to themselves. If being true to yourself means uniting the inside and the outside, aligning the su bjective processes and social performances, the first challenge is to find that self. The self becomes the origin to which authenticity refers and on the basis of which any claim to authenticity is evaluated.

We have come across several concepts of the self, all as critical responses to the idea of the self as a unitary omnipotent force in charge of itself and set out to conquer the world, an idea that has come under attack for its careless attitude towards society and environment (Madanipour, 2022a). The belief in the omnipotence and complete freedom of the human individual has been used as a licence for the excesses of individualism, colonizing the peoples of the world and exploiting natural resources to the edge of extinction. At the other end of the spectrum, there are those who argue that the self is non-existent, no more than a meeting point of blind natural forces, a mere accident without a consequence.

As the philosophers of mind have long argued, the sense of self is an inner, subjective process. Cognitive and neuroscientists have investigated the processes by which the brain and the bodywork interactively, but these scientists cannot engage in the subjective processes that construct the self. For psychoanalysts, this subjectivity is not fixed or stable, and it is always entangled in a process of repression and resistance, in a tense dialogue between the biological processes and the social order. Social psychologists have in turn emphasized the significance of society in the formation of this self, but this analysis makes it even harder to pinpoint the self as that clear foundation for authenticity. The self becomes dispersed and disintegrated through social pressures, norms, and habits, losing its distinctiveness, and no more than the reflection of a collective identity. The biological self, the psychological self, and the social self are all components of what we understand as the self. The search for authenticity leads us to the maxim Be True to Yourself, but we find this self to be a functional consciousness that is a workspace for the body, a fragmented subjectivity that is the scene of conflicting urges, and a reflexive, mediated process that locates us in our surrounding world.

Rather than a solid basis on which to construct the claim to authenticity, the self is a multi-layered, multifaceted, and continually changing process; while some of these layers are relatively stable, others change at faster rates. The self does exist, but not as a fixed entity, or a pure source of power detached from the world. Moreover, a simple unity of the inside and outside, of subjectivity and performance, and by extension the private and public spheres, is not possible or even desirable. The desire for authenticity comes from the human ability for self-reflection, and the desire for bridging the gap that this observation reveals between what they see from the outside and what they feel from the inside. Although both of these views are mediated through many social and historical layers, the existence of a gap between these two perspectives is felt by the person and by the others. Beyond this analysis of the self as a psychological state of mind or a social construct, we would need to analyse it as an active agent, which is the subject of the next chapter. Is it at the point of action that the self can be found and authenticity claimed?

This interdependence between the self and other is clearly reflected in the urban environment, where the relations of inside and outside, of public and private spaces are interdependent, rather than clear-cut, dichotomous entities. This suggests that authenticity is not to be found in a standalone inner subjectivity or a detached private space that keeps the others out, but in a dialogical process between the inside and outside spaces, and in the articulation of the boundaries and relationships between these different spaces.

# 3 Author

If authenticity is the quality of an object that shows its direct link with an origin, then one of the most important forms of that origin would be the power that produced that object. If we can show who produced the object, and people accept that producer to be qualified for doing so, one of the conditions of authenticity would have been met. In this chapter, we look at the source that is the basis of this form of authenticity, the author whose authority is recognized by others as being legitimate for producing the work. This chapter problematizes the power of authors in the making of four domains: the self, the text, the work of art, and the city. The question that this chapter raises is about cities, and whether such a source can be found for a city or any of its parts so that it can be declared an authentic city.

## Authenticity and authority

In English, French, and other European languages, there has long been an association between authentic and author (Oxford English Dictionary: author). Authentic and authenticity are derived from authent, an obsolete English word that was borrowed from Latin, where it meant author (OED: authent). The roots of authent go further back to the ancient Greek word *authentes*, which meant author and perpetrator, formed of a combination of self (auto) and to accomplish (*entes*) (OED: authent). The word *authentes* appeared to have three meanings in ancient Greece: 'murderer', 'master', and 'doer', leading to disputes about the relationship between these senses, and about their influence on the word's derivatives (Wolters, 2000). After all, this was a society in which writing, and the role of writers, could be a subject of dispute (Plato, 1997, *Phaedrus*). Nevertheless, what these three meanings of *authentes* seem to share is a reference to an actor with the power to undertake action, which could include dominating or even eliminating others. In a broader sense, authent indicates a relationship between agency and action, which finds expression in the terms author and authority.

The common meanings of authentic refer to a link between a creator and a creation. The creation may be an object, such as a legal document or a work of art, or it can be a state of being, such as everyday behaviour, personal character,

DOI: 10.4324/9781003451778-3

or a course of life. In the case of objects, the creator is often expected to be endowed with a power of making, a qualification that is recognized by others, an authority that is affirmable by society. This is exemplified by a person who is known to be an artist or an organization that is legitimately authorized to issue certain documents. As a result of the originators' common recognition, the work attributed to them is also recognized as authentic. If you have a passport or a higher educational degree, and you can show that your document was issued by the relevant government agency or a reputable university, then your document would be authenticated by virtue of its production by that official organization. Similarly, a book or a work of art is authenticated if the author or the artist who produced it can be identified.

In these cases, authenticity is proved by demonstrating a direct link between a document and its author, between a work and its originator. Being authentic means confirming the reliability of a relationship between a phenomenon and an origin; that is, a relationship between a name, statement, reproduction, event, or expression, and something or someone. The causal link between the creator and the created, therefore, needs to be demonstrated and accepted by others. This requires a process of judgement, an authentication. Any claim to authenticity depends on the visibility of that link, and on its evaluation and subsequent recognition and acceptability by others. This relationship between a creator and a creation, however, is more complex when it comes to the authenticity of a state of being, where many questions and doubts may emerge.

Through its links with an author, the notion of authenticity finds an identifiable, reliable source; and through the power of that source, an authentic person, behaviour, or object becomes empowered, socially recognized, and valued. The questions that should be raised immediately are about the power and legitimacy of the author: is the power of the author real or legitimate? What are the conditions that allow a person or institution to find that power of authorship, which gives them an authority, and an upper hand in their dealing with others? Can a work be completely attributable to a single author, or is the author no more than a symbolic origin? Is the author empowered through the act of making, or through the reception and recognition by others? What are the power dynamics of a concept like authenticity? In this chapter, we will look for some answers to these questions in several fields of authorship: the self, the text, the work of art, and the city.

## Authoring the self

The maxim Be True to Yourself has been interpreted as discovering an inner self and displaying it to others, thereby aligning the inside and outside, as discussed in the last chapter. When the outside is thought to be a true reflection of the inside, that person is identified as being authentic. However, the inside is multidimensional and ever-changing, and the situations in which we find ourselves are never the same. Therefore, the outside cannot be loyal to that dynamic interior and remain the same. To overcome this mismatch, and to ensure the inside-outside alignment, different masks are created and displayed to respond to these diversities and instabilities. The acts of self-discovery and inside–outside alignment show an actor at work, a performer engaged in an elaborate performance, an author who develops convincing narratives of the self, out of its many sides and dimensions, and displays these narratives on the outside, in appearance and actions and according to different situations, so that self-identity and self-presentation coincide. The audience may or may not believe the narrative, but they may expect an author to lie behind these performances and suspect a distance between the actor and the act.

### *Performing the self*

The self is not merely constructed cognitively through self-knowledge, but primarily interactively, engaged with the surrounding environment. We look at ourselves through the eyes of others and wish to be seen in a good light. As a result, from very early on in childhood we develop the skills of performance so that we can communicate with others and present ourselves in a positive way. As William James describes the social self in his 1890 book *The Principles of Psychology*, we wear different masks according to different roles that we play in society, with family, friends, strangers, and authorities; accordingly, we have 'as many different social selves as there are distinct *groups* of persons' whose opinion matters to us (James, quoted in Goffman, 1969:42). Erving Goffman (1969:74) argued that in a social encounter, we try to control the situation, and the definition of reality, by controlling the impression that the others have of us.

As such, social life becomes a performance, and intersubjective spaces are the stage of a theatre on which this drama unfolds. There is always a backstage, however, where the performers drop their masks. Meanwhile, the spectator/performer takes pleasure to see the moments in which the mask of the others slips, and their vulnerabilities are revealed in their 'naked unsocialized look' (Goffman, 1969:207). Goffman analyses the attempts to control the non-discursive elements of social encounters, such as gestures and manners, but the idea of the self also emerges discursively. The presence of others will always have an impact on what we think and do, and the self will not become so transparent that no distinction can be made between the inside and outside. Performance in response to the presence of others would always remain a primary ingredient of social encounters. As Zizek (2006:32) argues, reflecting Lacan, this performance may be a more authentic expression of a person than their inner subjectivity: 'the emotions that I perform through the mask (the false persona) that I adopt can in a strange way be more authentic and truthful than what I assume that I feel in myself'. The person's identity becomes equated with their performance, the self is equated with behaviour rather than subjectivity, an idea that has been widely applied in different fields (Butler, 1990).

Taking a selfie with a smartphone is a very common occurrence, and even if it may have felt somewhat awkward at some point in time, it has become a normal event across the world. The selfie may appear to be a snapshot of the self at one moment, but it is also the construction of an appearance, an exercise in authoring a self that is presented to oneself and to others. It is simultaneously a reflection of what we see and what we want to be seen, the capturing of a moment and the fabrication of a mask. Poor-quality selfies are deleted, and only those kept and shown that would show us in a good light, an appearance that would secure social recognition and approval. The selfie is not just a fleeting image in a technologically mediated mirror, but an authored narrative about a person in a particular situation, which has the inherent potential for presentation to others.

In everyday life, social masks are used when encountering others. But there are also moments in which these masks are put aside when the moments of heightened emotions can make it difficult to maintain social composure and the wearing of a presentable mask. On social media, much more control is usually available in the presentation of the self. The making of the self in social media shows how an image is carefully produced for presentation to others, with the help of ornamentation and all available performative tools, an image that is an exchangeable commodity in the marketplace. This is particularly the case when the number of likes on a post is directly related to the number of advertisements that it can generate, and hence the income that the maker of the post can earn. This is a form of commercialization of the self that is not seeking an original self but constructing a commodified one, which can be exchanged for money in global networks. Here too, however, there are moments in which the mask is slipped away when careless and hurried posts can lead to disastrous results for their makers, ranging from damage to the carefully constructed reputation to the complete loss of personal freedom and livelihood.

What is authored is not a whole self but a performance, showing a side of the self that is adaptive to specific circumstances. It is a narrative, an account of personal identity that links different fragments and polishes them into socially recognizable masks and performances, rather than being able to capture, control, and shape what lies behind the mask. If there is any authenticity to be found, it would be in the narrative, in the relationship between the mask and what lies behind it. We shall come back to this theme of performativity in Chapter 5, discussing purposeful performances in language, politics, and performing arts.

### *Autonomy and authenticity*

Concern about self-image and constructing a self in the hope of social recognition are not new. The ancient and medieval books and plays are full of instructions about how to behave in different situations, how children should be educated, and how to build character. The task of the parents, teachers, priests, and others in positions of authority, was to shape the character of the child and guide the behaviour of the adult so that individuals could conform to the norms of society. While these norms, and the laws that supported them, differed widely in different societies and periods of history, what was shared between them was a demand for conformity, deemed essential for the maintenance of the social fabric. You were responsible for your actions and therefore encouraged to be the author of yourself, but in a way that would be expected of you. You ought to be the author of yourself, as long as this self was clearly delimited and controlled along the socially accepted lines; in other words, the main author of the self was the society and you acted as its agent. This can be seen in various schools of thought, from Socrates to Stoics, in various religious traditions across different continents from Abrahamic religions to Confucius, and in the everyday life of most communities around the world and throughout history. However, this sociocentric view of the world appeared to start changing with the end of the Middle Ages in Europe (Coleman, 1996; Taylor, 1989).

The post-Renaissance Europe asserted the significance of humans. The theocentric worldview of the medieval period gave way to an anthropocentric perspective in which humans now stood at the centre of the world. This is widely represented in paintings, sculpture, and the organization of space (Pevsner, 1963). It is shown by the concerns about self-image, as reflected in memoirs, autobiographies, and self-portraits, where reflection on the self becomes an acceptable practice. Michel de Montaigne (2004:1) published his *Essays* in 1580 for the 'private benefit of friends and kinsmen', to present 'some traits of my character and of my humours'. The skills of the painter would be directed inwards, ranging from the confident self-image of Albrecht Dürer to the emotional nuances throughout a lifetime as captured by the self-portraits of Rembrandt, all constructing a narrative of the individual self and articulating the way that the self is presented to others. It is a self that is authored through the act of artistic self-expression.

A philosophical landmark in this transition to anthropocentrism was laid by Descartes who, writing in the first half of the seventeenth century, looked for a solid foundation for knowledge in the human mind, as expressed in his famous phrase, 'I think, therefore I am' (Descartes, 1968:53, Discourse 4). In circumstances of being 'able to live as solitary and withdrawn as I would in the most remote of deserts' (p.52, Discourse 3), Descartes aimed at abandoning 'all the opinions I had accepted' and 'replace them afterwards either by better ones, or even by the same, once I had adjusted them by the plumb-line of reason' (p.37, Discourse 2). By relying on reason, rather than 'custom and example' (p.39, Discourse 2), he was convinced that humans could find a practical philosophy with which to 'make ourselves, as it were, masters and possessors of nature' (p.78, Discourse 6). Rather than being fearful of natural forces, relying on supernatural powers, or being bound by tradition and custom, humans were now thought to have the power of reshaping their own minds and transforming their world.

A radicalization of this notion was made by Kant who, in his Copernican revolution a century and a half after Descartes, located human subjectivity at the centre of knowledge and action. He split the world into what is and what appears to us, arguing that we can only understand that appearance, and this understanding reflects our limited cognitive powers (Kant, 1987). Kant invited his readers to 'Dare to know', and 'Have courage to use your own reason', which he identified as 'the motto of enlightenment' (Kant, 1995:1). Rather than being under the tutelage of others who would act as their guardians, his readers were encouraged to become independent:

> If I have a book which understands for me, a pastor who has a conscience for me, a physician who decides my diet, and so forth, I need not trouble myself. I need not think, if I can only pay – others will readily undertake the irksome work for me.
>
> (Kant, 1995:1)

For a public to become able to think for itself, what was needed was freedom within a framework: 'if only freedom is granted, enlightenment is almost sure to follow' (p.2). Kant advocated autonomy and freedom of choice framed by universal moral laws that were intuitively rational (Kant, 1987; Kisner and Noller, 2020).

For others, the self was not necessarily taken to be a source of rational thinking, but a reference for a person's unique identity. Rousseau (1783:1) started his *Confessions* by writing, 'I am undertaking a work which has no example, and whose execution will have no imitator. I mean to lay open to my fellow-mortals a man just as nature wrought him; and this man is myself'. As the title of his book suggests, self-reflection and confession are rooted in the Christian tradition, but now it is no longer a private event, but secularized and presented to the public. Rousseau's *Confessions* shows searching inwards for a source that could guide him in his life. However, this introspection looks for something further and deeper than what the rationalists had found. It was not finding a higher self, as the mystical traditions had searched for, but the inner passions and tensions. As a pioneer of romanticism, Rousseau is also a pioneer of the idea of authenticity (Varga and Guignon, 2023).

Therefore, authenticity should be distinguished from autonomy, despite their overlaps and similarities (Varga and Guignon, 2023). While autonomy and authenticity both promoted the independence of the self, their image of the self was not the same. For Descartes, Kant and the enlightenment tradition, the supremacy of reason was an essential ingredient of autonomy, whereas the romantics looked beyond rationality to find a deeper foundation within the self. The critical response to the emergence of the urban industrial world in the eighteenth and nineteenth centuries was that the rational self of the enlightenment was too narrow and restrictive.

In autonomy and authenticity, the author of the self and the target of authorship were not the same.

Another important distinction that needs to be made is between different meanings of individuality, as they emerge out of the evolution of the concept and the transformation of its context. The assertion of individuality in the early modern age is a critical response to the religious laws and restrictions of the medieval period, but those who made this assertion could live with the absolute power of monarchies of their time. The assertion of individuality also emerges in critical reaction to the rising centralization of power in absolute monarchies, which triggered the republican revolution in France and the United States, and the liberalization of monarchies. The assertion of individuality is also made in reaction to the restrictions made by industrial capitalism and its urban mass culture, the tight bureaucratic controls of the socialist states, and the extended anonymity and sense of insignificance that are associated with globalized digital economy. While all are cries for freedom in the face of powerful forces, the social and historical conditions, the nature of restrictions, and their images of the free individual are widely different, as are their ideas about autonomy and authenticity.

### *Creating an authentic self*

The concept of authenticity became the cornerstone of existentialism, a philosophical tradition that appeared in the first half of the twentieth century in the work of Heidegger and Jaspers, and rooted in the early nineteenth-century German thought, which emerged as a sharp criticism of the reigning Hegelianism and rationalism (Tillich, 1944). It drew on Rousseau and the romantic tradition, on Kierkegaard who rejected institutionalized religion and searched for an individual basis for his belief, on Nietzsche who advocated the freedom of the individual from social constraints, and on Husserl who radicalized Descartes by developing a phenomenological method of understanding. The arts became a model for life, and the subject was to be an artist at work, creating the self and the course of life like a work of art, as a unique, personal creation (Figure 3.1).

This line of thought revived the scholastic distinction between essence and existence, arguing that reality is what people immediately experience in their living, rather than in an essence that has resulted from rational thinking. Rationalist thinking had equated reality with the object of reason or thought, while existentialists were among a wide range of thinkers who emphasized the primacy of immediate experience in grasping reality (Tillich, 1944: 44–45). Heidegger drew on Husserl, whose phenomenology had shifted the emphasis from objects to the mind, and he anchored the meaning of existence in concrete experience, through his concept of Dasein, which meant 'being in a special place', or 'being *da* or "there"' (p.47).

*Figure 3.1* The idea of endless potentialities in self-creation is a modern concept (Newcastle, UK)

According to Heidegger, however, this concrete existence needs to be cleansed from the contaminations of modern industrial society. Therefore, the authentic self is a human being who is 'there' but is not polluted by 'they':

> The Self of everyday Dasein is the *they-self*, which we distinguish from the *authentic Self*-that is, from the Self which has been taken hold of in its own way [eigens ergriffenen]. As they-self, the particular Dasein has been *dispersed* into the "they", and must first find itself.
>
> (Heidegger, 1962:167, I.4, 129)

This detachment from the outside and withdrawal into the inside in search of a solid foundation is reminiscent of the Cartesian approach but without its praise for reason. However, the possibility of discovering a self-identity that is detached from society may prove impossible. 'To be authentic, Dasein must be triply alienated, alienated from everyday concerns, from the public world, and from public interpretations of everyday concerns' (Zuidervaart, 2007:88). Adorno (1973) agrees with the diagnosis of alienation and dispersion in modern society, but he is critical of Heidegger's response. He attacks what he calls 'The Jargon of Authenticity' (Adorno, 1973), criticizing Heidegger's inwardness and arguing that this isolated self would be as diminished as the dispersed self:

> He does not think of the connection between the large urban center, of high capitalism and that dispersion which was noted by Georg Simmel and already felt by Baudelaire. Whatever remains solely by itself, as one's authentic existence, becomes no less impoverished than that which dissolves into situations.
>
> (Adorno, 1973:77–78)

Adorno argues that the reconciliation between the inner and outer worlds, which Hegel had hoped for, would be in this way postponed forever. In a world where alienation has already become the norm, Adorno argues, 'it has become unnecessary to advocate alienation, since the latter is in power anyway' (Adorno, 1973:78).

Sartre (2003) drew on Heidegger to develop his own version of existentialism. In normal circumstances, we are inauthentic, and it requires self-awareness and taking action to become authentic. Authenticity is a self-recovery of being from inauthenticity, which is 'my ordinary state in so far as I have not realized my conversion to authenticity' (Sartre, 2003:246). The motivation for this conversion is the recognition of our contingency (p.80). When this contingency is acknowledged and embraced, the path towards authenticity is opened. At the moment that 'I launch out toward death with a resolute-decision..., I reveal myself to myself in authenticity, and I raise others along with myself toward the authentic' (p.246). This will facilitate freedom from 'everyday banality in order to attain the irreplaceable uniqueness of the person' (p.534).

The notion of authoring the self can be found even in the critics of existentialism, even if they do not use the concept of authenticity. After admitting in an interview that he had been influenced by Heidegger (Martin, 1988:12), Foucault hints at his idea of the authorship of the self: 'I don't feel that it is necessary to know exactly what I am. The main interest in life and work is to become someone else that you were not in the beginning' (Martin, 1988:9). In his analysis of ancient Greece and Rome, Foucault questions the adequacy of social determination for self-satisfaction. 'From the viewpoint of the relation to the self, the social and political identifications do not function as authentic marks of a mode of being; they are extrinsic, artificial, and unfounded signs' (Foucault, 1986:94). Instead, he describes 'the practices of the self', which were recommended by the Stoics, which included the care and cultivation of the self (Foucault, 1986; Hadot, 1995). For the Greeks, the precept 'to be concerned with oneself' was 'one of the main principles of cities, one of the main rules for social and personal conduct and the art of life' (Foucault, 1988:19). Priority was given to caring for the self over self-knowledge, while this changes in the Christian, and later in the modern period, when the emphasis has been on self-knowledge, reversing the ancient morality (Foucault, 1988:22).

The version of existentialism that was developed in France in the mid-twentieth century drew on the German tradition but also on a Cartesian idea of the free subject (Marcuse, 1948). According to Sartre, existentialism is based on the idea that 'existence precedes and perpetually creates the essence' (quoted in Marcuse, 1948:311). However, Marcuse (1948:309) argues, while the Cartesian subject was conscious of its power and 'faced an objective world which rewarded calculation, conquest, and domination', this new subject becomes 'absurd and its world void of purpose and hope'. According to Sartre, in this absurd and meaningless world, the subject is free to create itself and create meaning for its life. Human beings are free and responsible for creating their own lives; they are therefore 'nothing but self-creation', or 'permanent realization' of their possibilities (Marcuse, 1948:314). Their existence is the realization of a project, 'freely designed and freely executed' by themselves (p.315). Marcuse criticizes the non-historical character of this conceptual model, as an abstract model outside time and space.

Like Adorno (2004:49–50), who criticized Sartre's existentialism as 'decisionism' and harking back to 'the old idealistic category of the free act of the subject', Marcuse rejects the self-centred nature of this way of thinking, which limited the understanding of the other. Against the idea of humans' absolute freedom, Marcuse argues that humans are 'determined' by their 'specific socio-historical situations, which in turn determines the scope and content' of their liberty and the range of their 'choice' (Marcuse, 1948:320). Marcuse accuses existentialism to be a handmaiden of capitalism: 'Behind the nihilistic language of Existentialism lurks the ideology of free competition, free initiative, and equal opportunity' (p.323).

This critique by Adorno and Marcuse draws on social determination to reject the idea of self-determination. It rejects the belief in the ability of the human subject to become autonomous or authentic, to be an independent author of the self that shapes its own pathways through life, arguing that the self is always shaped and framed by its environment. According to their Marxian, materialist critique, the notion of being the author of oneself is idealistic, and authenticity is no more than jargon. In a framework of determinism, the maxim being true to yourself, and the ideas of agency and personal responsibility, lose their meaning, as individuals become the results of their circumstances, rather than their own conscious decisions. However, the problem with this emphasis on social determinism is the neglect of human beings as having any meaningful agency in making decisions for their own life. It shows a movement from one extreme to another, from total self-determinism to total social determinism, both suffering from reductionism. An alternative view is to understand the conditions of possibility for the emergence of a phenomenon, but acknowledging that it can go beyond these conditions, rather than being determined by them (Badiou, 2005).

Concern about the idea of authoring the self is giving the impression that it is a normal process of constructing something tangible and stable. As we saw in the last chapter, the self is not static nor coherent, but has many evolving dimensions engaged in diverse situations; therefore, the idea of making it is not a single process of decision making and authorship. What are the dynamics of authoring a text, where the name might be more appropriately used?

## Authoring the text

The tension between social determination and self-determination finds manifestation in the role of the author. If authenticity is based on tracing back a text to an author, what can authenticity mean if the author is exerting power and authority, or if the identity and authority of the author are lost? Authors are presumed to produce text, and in doing so they exert a degree of authority in their relations with their readers and others. In many situations, texts may be partly evaluated through the author's identity. This is why the work of famous authors could automatically be given a high value, even if the work itself is of moderate quality. Today, owing to educational and technological developments, the number of authors may have proliferated to unprecedented levels, and subsequently, the authors may not have the power that they used to enjoy through the limited range of the media in partly literate societies of the past. But the power of the author is still strong when compared with those who have no voice (Rancière, 2013). At the same time, what happens if the significance of the author is questioned, and the reader is thought to hold the power of reading and interpreting the text?

In physical sciences, the power of the author has long been diminished, and scientific knowledge is expected to stand on its own, evaluated through critical examination by other scientists, rather than believed because of its author's authority. This has been deployed in the tradition of blind refereeing of scientific articles, in which the identity of the author is hidden from the reviewer. What gives this knowledge its validity is its replicability and its openness to critique and revision by others. While scientists may become well known in their fields, and this may give them some advantages in receiving grants, gaining positions of authority, prizes, and wider social recognition, they cannot rely on this authority to claim new knowledge. It is the approval of the scientific community that gives value to a discourse, rather than the authority of its author. The concept of authenticity would require establishing a link between the text and a source; in this context, if the concept is at all applicable, that source is formed of a set of practices, relationships, and paradigms that constitute a community of practitioners. The author's role is not denied, but it finds reliability only within a particular framework. This diminishing role of the author vis a vis a structure of reliability can also be seen in human sciences, where it has been far more controversial.

As Michel Foucault (1981:58) observes, since the seventeenth century the function of the author has been eroded in sciences but has grown stronger in literature. Roland Barthes (1977:142–143) argues that the 'author is a modern figure', coming to prominence alongside the growth of attention to the significance of humans, and in line with positivism, as the ideology of capitalism, that emphasizes the 'person' of the author. The common interpretations of the works of writers, musicians, and painters are 'tyrannically centred on the author' (p.143). Critical voices, therefore, set out to dethrone the author. In a famous essay in 1967, entitled The Death of the Author, Roland Barthes argued that it is the reader's interpretation, rather than the author's character or intention, that gives meaning to the text. Barthes suggests that 'writing is the destruction of every voice, of every point of origin': as soon as something is written, a disconnection occurs between the source and the work, which marks the death of the author (p.142). In other words, 'it is language which speaks, not the author' (p.143), and language 'ceaselessly calls into question all origins' (p.146).

What is written is 'a tissue of quotations drawn from the innumerable centres of culture', it is 'a multi-dimensional space', in which different fragments 'blend and clash', and the author's 'only power is to mix writings' (Barthes, 1977:146). It is the reader who brings those fragments together and gives unity to the text. In other words, 'the birth of the reader must be at the cost of the death of the Author' (p.148). Rather than establishing the authenticity of a text by tracing it to the authority of an author, the text is linked to language as a framing structure, diminishing the significance of the author in an act of political and literary critique. This is a challenge to the figures of authority and established orders; it refuses to fix meaning and sees itself as a revolutionary 'anti-theological activity' (p.147).

Michel Foucault (1981) raises a similar critique of the author, describing the author as no more than a function in the production of discourse. Foucault asserts that this is not 'to deny the existence of the individual who writes and invents', which 'would of course be absurd' (p.59), but to see the author as part of larger procedures of power. In his inaugural lecture at the Collège de France in 1970, he put forward a general hypothesis that in every society discourse is linked to procedures of power, which control and delimit discourse. One of these procedures is the author-function, which Foucault sees as a principle that gives order and coherence to different parts of discourse and provides an origin for their meaning. The function of the author is a process of inclusion and exclusion: 'what he writes and what he does not write', a function that 'he receives it from his epoch, or as he modifies it in his turn' (pp.58–59). In search of a name for his ideas, Foucault asserts that 'all this is structuralism' (p.73), a label that he later denounced.

In these critiques, Barthes and Foucault reject seeing the author as a point of origin, which was the hallmark of existentialism, and question the author's freedom to create meaning, a freedom which is commonly presumed since the rise of modern individualism. It is a critique of agency in the name of structure, paying attention to the overall conditions in which actions are taken, rather than the primacy of the author as the generator of meaning and action. There is a similar socio-centrism in Lacan, where the symbolic order reigns, and the individual subjectivity is where 'the big Other, the anonymous symbolic order, gets subjectivized' (Zizek, 2006:41).

The implications are that, if the author is a mere cog in a linguistic system or a function in a structure of power, there will be no source of authenticity and originality to be found. The concept of authenticity loses its meaning altogether, as meaning is always diffuse, a part of a never-ending chain of signification. As Barthes puts it, the author's 'only power is to mix writing', drawn from 'a ready-formed dictionary, its words only explainable through other words, and so on indefinitely', while 'the book itself is only a tissue of signs, an imitation that is lost, infinitely deferred' (Barthes, 1977:146).

In these critiques, the existence of the author is not denied. It is the agency of the author as an origin of meaning and action that is denied, and subsequently the possibility of inward search for an authentic self, and the ability to construct a unique and independent course for life, disappear. However, in this denial of the authority of the author, there is a shift of power to the reader. Should the search for authenticity be transferred to the realm of readers, and can every reading be considered an authentic reading? The answer that Gadamer (2004), a prominent follower of Heidegger, provides to this challenge is that there is a halfway house between the author and the reader, where a 'fusion of horizons' takes place, and where meaning is created dialogically. Rather than a tug of war between the author and the reader, they are both acknowledged as having a role in the exchange and sharing of ideas.

## Artists and artworks

Aristotle (1991) considers art as the process of making contingent objects; these objects do not exist by necessity, and so they are 'capable of either being or not being'; therefore their 'origin is in the maker and not in the thing made' (Aristotle, *Nicomachean Ethics*, VI, §4). Therefore, a work of art is authenticated through its link with an artist, the one who is the author of the work. The artist is identified as the origin of the work, and this gives authority to the artist and authenticity to the work (Figure 3.2). This explicit and significant relationship between the artwork and the artist is particularly emphasized after the European Renaissance, radicalized since the eighteenth century. In the medieval period, the separation of art from everyday life, and the celebration of the artist, did not exist in the same way. The arts and crafts were an integral part of religious performance, aristocratic lifestyle, and everyday practices of society. It was with the start of the Renaissance in Europe that named artists became well-known and their work stood out for their artistic value. Alongside the rise in humanism, the significance of the self, and asserting the centrality of the author for the text, there is a growing attention to the authority of visual and spatial artists.

This trend reaches new heights in the eighteenth century with the cult of genius. The key contribution that Kant made to the theory of fine arts was

*Figure 3.2* Artworks are authenticated by reference to the artists. Are there other criteria with which their authenticity is assessed? (Barcelona, Spain)

his analysis of genius (Pluhar, 1987:xxiv). Alongside his argument for the necessity of personal autonomy based on rationality, Kant argued for the necessity of personal talent in making art, thereby radicalizing the role of the individual artist in the production of the artwork. He argues that 'fine art is possible only as the product of genius', as 'Genius is the talent (natural endowment) that gives the rule to art' (Kant, 1987:174–175; §46). According to Kant, judging the fine arts should be based on taste, but producing the fine arts requires genius. Genius needs to be cultivated by taste, but it is a talent that is not constrained by rules; instead, it produces works that become examples that set the rules for others to follow, but this cannot be taught or communicated.

As Kant puts it, 'That is why, if an author owes a product to his genius, he himself does not know how he came by the ideas for it', and therefore unable to produce them at will or by plan, and unable to communicate his procedures to others so that they can produce similar works (Kant, 1987:175; §46). In other words, 'the rule must be abstracted from what the artist has done, i.e., from the product, which others may use to test their own talent' (p.177, §47). In this way, the work of art is traced back to an inner talent that is only found in some individuals; this talent is not explainable but can only be manifested in its product and can only be explained after the event. What is emphasized in this formula is the idea of genius and the centrality of individual artist, who is not required or able to reflect or rationally explain that talent and how it produces the artwork.

In critical reaction to the idea of isolated genius, Hegel (1975, I, iii, C) locates the artist in continuous interaction with the world, rather than being detached from it. He does not deny the existence of genius: 'Genius is the general ability for the true production of a work of art, as well as the energy to elaborate and complete it' (Hegel, 1975:283). But he sees it as part of a process of relating the inner and outer worlds. 'Without circumspection, discrimination, and criticism the artist cannot master any subject-matter which he is to configurate, and it is silly to believe that the genuine artist does not know what he is doing' (p.283). Hegel agrees that the work of art has been shaped by 'creative subjectivity' (p.280), but this has not developed in isolation, and 'the artist is not relegated to what he has manufactured by his own imagination but has to ...enter reality itself' (p.281). Beyond an inborn talent and a knowledge of the inner life and the soul's passions and aims (p.282), this requires immersion in the world, as the artist 'must have seen much, heard much, and retained much' (p.281), developed technical dexterity to give external shape to inner imagination (p.285), and be inspired by the surrounding world (p.288).

The image of the artist as someone who is above the rules and norms of society and is primarily devoted to self-discovery and self-realization finds its full expression in Friedrich Nietzsche's writings. The subtitle of his book *Ecce Homo* (Nietzsche, 2007) is: 'How one becomes what you are', and elsewhere he asks, '*What does your conscience say?* – "You should become who you are."' (Nietzsche, 2001:152, No.270). Rejecting Kant's universalist ethics, Nietzsche (2001:189, No.335) puts forward a voluntarist proposal: 'We, however, *want to become who we are* – human beings who are new, unique, incomparable, who give themselves laws, who create themselves!'. Life is to be modelled after a distinctive work of art that needs to be created by the self-author through sheer will, rather than through conformity to established norms. Nevertheless, this self-creating artist appears to be like Kant's genius, not being constrained by norms and laws but involved in creating them, a genius who is no longer limited to the realm of aesthetics. There is a direct link between the Kantian aesthetics of genius and the Nietzschean ethics of Superman.

The Hegelian line of critique shows an attempt to explain how creativity is nourished and developed, rather than putting an emphasis on the significance of individuality and authorship, a debate that puts the mediation of the contexts and technical abilities against the idea of an unexplainable inner power of creativity. It is a precursor to the tension between the author and the text, as we saw earlier, reflecting broader social and political divergences. Is the source of authenticity a single, highly creative author, a genius, or is it a series of conditions that enable that creativity to flourish and find expression? Is it the outcome of a single subjectivity or a reflection of complex conditions of possibility? These debates also reflect major controversies about the meaning and role of art, as well as the function of art criticism and art history (Iversen and Melville, 2010).

The question of the author leads to further questions about whether an artwork primarily reflects the artist's subjectivity or is an expression of a culture and society. What does art do? Does it generate access to an inner power of creativity, a self that is unknown even to the artist? Can the artwork be understood and explained through the psychology and biography of the artists, or through the social and historical contexts in which they live? Can individuals act outside these contexts, or are they the slaves of these contexts, or is there a scope for choice even within the tightest of conditions? These questions have direct implications for the analysis of authenticity, as they guide the search in one or another direction. If authenticity is judged on the basis of a link with an origin, is that origin a single author, a solitary artist, or a whole host of actions and conditions that form that origin? If the latter, a further question is whether a clearly identifiable origin can be found in such complexity, and thereby whether authenticity may be discoverable or can exist at all. These questions also find expression in the design and development of cities.

## Designers and the city

The search for authenticity in an author may lead to the designers, who are seen as the authors of the city. What is the extent of their power, and how far can city-making be associated with single designers and single sources of authority?

### *Nameless makers*

Before the modern emphasis on individual authority, the craftsmen who made objects in traditional ways were not seen as the origin of a work, but as operators of a long line of skills and knowledge that was inherited from the past. Their products, which may be considered as works of art today, were functional, contextualized objects, playing a specified social function. These objects were not produced or seen as expressions of individuality, but a reflection of social conventions and needs, which were creatively shaped in the same or slightly new ways. The importance of individual talent and skill was always recognized, but as part of a social texture, and not as a heroic genius that stood above the rest and was free from all social considerations. In these cases, the origin that determines the case for authenticity could be a tradition, a technique, or a material, but not a named author. The artwork would not be signed, as the artists-craftsmen were expected to contribute to a way of life, rather than projecting their own personality and significance.

In the urban environment, this is shown in the vernacular architecture, which has been the main form of urbanism in much the world and throughout

*Figure 3.3* Much of the world's urban environment throughout history has been made by nameless makers (Meybod, Iran)

history (Figure 3.3). The city is not built by a single company or designed by a named architect, but it is a gradual, collective creation that often bears no particular person's name. The relations of power are surely at work, and there are those individuals in powerful positions that can influence the shape and conditions of the city. But a village, a town or a city is always a historical product in the making, expanding or contracting through incremental steps. In such circumstances, the significance of a single authority is often lost in the complexity and multiplicity of historical actors and processes. When we see a city, we see a surface behind which lies the historical accumulation of material conditions and social relations, intertwined and multilayered, where any single authority's role may be limited in scope and scale. In all periods of history, cities or some of their significant parts may be attributed to the authority of a single political leader or designer, but the process of city-making is always more complex, multidimensional, and multiagency.

In a famous exhibition entitled Architecture Without Architects at the Museum of Modern Art in New York, Bernard Rudofsky introduced many examples of vernacular architecture from around the world. The exhibition was an attempt 'to break down our narrow concepts of the art of building by introducing the unfamiliar world of nonpedigreed architecture' (Rudofsky, 1964:2). For Rudofsky, this unfamiliar architecture was 'so little known that we don't even have a name for it. For want of a generic label, we shall call it vernacular, anonymous, spontaneous, indigenous, rural, as the case may be' (Rudofsky, 1964:2). What these works shared was that they had not been designed by well-known, university-educated, professional architects, but by craftsmen relying on the accumulated knowledge of local communities. The aim of the exhibition was broadening the scope of architecture to different geographies and histories beyond the Western-centred history of architecture. This was a direct critique of modernist architecture, at a time when the voices of dissent against it had started to rise. As Rudofsky (1964:2) wrote, 'Vernacular architecture does not go through fashion cycles. It is nearly immutable, indeed, unimprovable, since it serves its purpose to perfection. As a rule, the origin of indigenous building forms and construction methods is lost in the distant past'.

Rather than a person, the origin of building forms and construction methods is found in a process of evolution, where trial and error is repeated and adjusted in response to the needs and circumstances. This undermines the modern emphasis on the notion of architect as a genius, innovating artist-engineer, and gives up the hope for finding a point in time, a special type, or a specific hero as the origin of architecture. Rudofsky quotes Darwin who had observed animals that were involved in constructing their habitat. As Darwin (1871:53) wrote, 'The anthropomorphous apes, guided probably by instinct, build for themselves temporary platforms', orangs cover themselves at night with leaves, and baboons protect themselves from the heat of the sun by straw mats, all indicating the possibility of 'a voluntary and conscious act' at work. Darwin (1871:53) concludes that, 'In these latter habits, we probably see the first steps towards some of the simpler arts; namely rude architecture and dress, as they arose amongst the early progenitors of man'.

For an investigation into authenticity, this suggests that the origin of architecture would go deep into the process of evolution, rather than merely starting with the construction of the hut, as it has been frequently assumed. If finding an origin is necessary for accounting for the authenticity of something, then this origin would be very difficult to find in the construction of our habitat. In the same way, the origins of the towns and villages would go much further back before the spread of agriculture and the start of settled life. Such origins would lie in the notion of humans as social animals, hence the formation of small bands of human foragers living together, with similarities to how many animal species also live in groups. Rather than a single origin found in a single author, time, or place, the origins of human habitat are deep, long, and lost in the mists of time. If finding this origin is not easy in the process of evolution, it is also difficult to find the complexity of contemporary urban development.

The complexity of the urban development process, and the diversity and multiplicity of the agents at work, would make this task difficult, if not impossible. The industrial workers, who produce everyday objects and build cities, based on designs that have long been at work, with little contribution from a known designer, are the unknown and unknowable makers of objects. Modernist rationalist planners thought that they could know, predict, and control the future of the city, shaping it in the way they wished. In doing so, they integrated the process of urban design and planning into the bureaucratic processes of municipalities, making them more impersonal than ever before (Madanipour, 2015; Perloff, 1956).

### *Unity of political and aesthetic power*

The notion of authenticity is closely related to aesthetics (Taylor, 1991:63). Some of its early romantic advocates aimed at the void left by religion to be filled by art. As the artist created an original work of art, every individual was to make his or her life a work of art, original and unique. Rather than art imitating life, it was argued that life should imitate art, and everyone should be involved in the lifelong creative task of *poiesis*, making their own life according to artistic standards. This would avoid the dry rational uniformity, or the urban social conformity, generating an unprecedented aesthetic richness in society. This attention to the self as a work of art was also evident in the way cities were reshaped. It was argued that the city needed to be made like a work of art, to be improved (Laugier, 1977), according to the artistic standards (Sitte, 1986), which would turn the city into a collection of beautiful and meaningful places. Over centuries, European cities competed for building the most significant monuments, organizing their spaces in the most aesthetically pleasing arrangement. However, the aesthetic approach to the city, which followed the principles of originality and authenticity, was ousted by the arrival of modernist functionalism and mass construction technologies. This was a different aesthetic sensibility, creating a different form of originality and authenticity in the city.

The idea of authoring a city becomes popular during the Renaissance in Europe, alongside the rise of humanism and the growing concentration of political power. The designer, as the named individual engaged in purposefully and tastefully shaping an object, becomes distinguished from the nameless craftsmen of the past. It is thought that if a city is designed by a single designer, it would be more rational and better organized. Medieval cities are considered to be haphazard and badly built, and what is needed is reshaping them as a newly ordered entity. This sentiment is repeated in the twentieth century towards the nineteenth-century industrial city, a desire for a rational reorganization of the city which becomes the foundation of modernist urban design and planning. In much of the urbanizing world in the twentieth and twenty-first centuries, this desire for the 'creative destruction' (Schumpeter, 2003) of the old is still prevalent. As such, a continuous trend of trying to author the city along rational lines runs through six centuries and spreads throughout the world. But can the city be authored in this way, and can this authoring generate authenticity by virtue of tracing the city back to a single author? (Figure 3.4).

*Figure 3.4* Can reliance on a single, creative designer be sufficient for the making of an authentic city? (Milan, Italy)

Authoring the city is directly related to the presence of a strong political authority. The authority of an originator may have been invested in a powerful mythical or real figure, rather than in a designer. An example is the myth of twin brothers Romulus and Remus, who decide to build the city of Rome but their disagreement about its location is resolved by the murder of the latter, while Romulus goes on to establish the capital of a powerful kingdom named after himself (Potter, 2009). In recorded history, crediting a king for the foundation of a city is also common, as the famous example of Baghdad shows, which is attributed to the caliph Al-Mansur who is said to have personally selected the site and founded the capital of the Abbasid empire in 762 CE (Micheau, 2015). After searching for an appropriate site for a new capital, Al-Mansur declared: 'This is a good place for an army camp' and 'He sketched a plan of the city and put an army commander in charge of each quarter' (Tabari, 1995:238).

In the mythical story of Rome, events have been transformed into a symbolic narrative, as exemplified in the image of the wolf that nourished the twin brothers, which became a symbol of the city. In the recorded history of Baghdad, symbolism is not far, and as reported by a tenth-century historian, there are different versions of the story of the foundation, which show the involvement of others in finding the site and designing the city

(Tabari, 1995:237–252). The army camp that became the city was not built in a void, but it was the suburb of an existing metropolis, the old capital of the Persian empire, Ctesiphon, which had fallen in defeat. In both cases of mythological and recorded history, a complex history of urban development is reduced to a simple story and a single attribution, in which one powerful figure is given full credit for the creation of a metropolis. The mythology of a single origin allows history to start from a specific date, revolving around the will of a single person, disregarding the past and the many others who were involved in creating, inhabiting, and transforming the city. It facilitates telling powerful stories that capture the imagination and reinforce the city's reputation. It is a simple formula in which a phenomenon is explained and given identity by linking it to the power of an authority.

After the Renaissance, when absolute monarchies started to emerge in Europe, this story of a single powerful origin for a city became widespread in the design of ideal, rational cities, where the power of the ruler and the authority of the designer are married in the name of rationality and expressed in the form of geometrical regularity. This is exemplified in the work of Filarete (1965), an architect and theorist who, in his *Treatise on Architecture*, which he wrote between 1461 and 1464 in Milan, proposed the first star-shaped ideal city of the Renaissance (Madanipour, 2022b). According to Pevsner (1963:185), this was 'the first wholly symmetrical town plan in Western history'. Filarete named his ideal city Sforzinda, after the powerful Duke of Milan, manifesting the direct link between the concentration of power in the hands of a political ruler and the creation of urban space by a single designer.

Filarete's ideal city design combines humanism and naturalism, thereby seeking harmony, proportion, and order, which are to be found in the human body and in mathematics. The human body, and particularly the human head, as perfectly shaped by God, was to be the model of the city and an anthropomorphic world. The head was 'the most noble and most beautiful member', determining the proportions of the body: 'if the arms are opened and the hands extended, [the man] will be nine heads in either direction' (Filarete, 1965:8). The human head was that single source of authority in determining the shape of the body, as the absolute ruler was expected to shape society. A well-ordered city was to be designed by a single designer and ruled by a single ruler (Madanipour, 2014). This anticipates Machiavelli (1950) who, a generation later, argued for the necessity of having a single source for legal and political design. It also anticipates Descartes who, two centuries later argued for cities to be designed by a single designer. However, while for Filarete, the point of reference for a rational world was the human body, for Descartes it was the human mind (Madanipour, 2022b).

City design was the model on which René Descartes (1968:37, Discourse 2) constructed his own philosophy, along the metaphoric 'plumb-line of reason'. By observing the cities of the past, and imagining the rational cities of the future, he deduced the idea of abandoning the inherited knowledge and looking inside his own mind in search of certainty (pp.36–37). He was dismissive of the historical cities and their gradual piecemeal and apparently chaotic growth, looking for a rational order that could only be achieved by a single designer. He was a mathematician looking for a systematic order, and so the non-mathematical arrangement of historical urban space disturbed him, appearing to be disordered and irrational. The charm of historical cities, which later was much praised by the nineteenth-century romantics, by the critics of twentieth-century modernism, and by masses of tourists in our time, was considered to be no more than the outdated disorder. There is a strong aesthetic sense in this rejection of visual and spatial disorder by the father of modern rationalist philosophy. As he wrote (Descartes, 1968:35),

> there is less perfection in works composed of several separate pieces and made by different masters, than in those at which only one person has worked. So it is that one sees that buildings undertaken and completed by a single architect are usually more beautiful and better ordered than those that several architects have tried to put into shape, making use of old walls which were built for other purposes. So it is that these old cities which, originally only villages, have become, through the passage of time, great towns, are usually so badly proportioned in comparison with those orderly towns which an engineer designs at will on some plain that, although the buildings, taken separately, often display as much art as those of the planned towns or even more, nevertheless, seeing how they are placed, with a big one here, a small one there, and how they cause the streets to bend and be at different levels, one has the impression that they are more the product of chance than that of a human will operating according to reason.

However, Descartes is aware that it is not easy to turn an entire city into a blank canvas on which a completely new rational order can be built. Therefore, he adjusts and rescales his expectations:

> we have no example of people demolishing all the houses in a town for the sole purpose of rebuilding them in a different way to make the streets more beautiful; but one does see many people knock their own in order to rebuild them.
>
> (pp.36–37)

He follows this example to knock down his own mental house and rebuild it in a rational way, so he examines his own beliefs, convinced that this would help him 'succeed in ordering my life much better than if I built only on old foundations' (p.37). Future generations do not constrain their desire for a rational order to single houses.

The search for constructing a well-ordered, rationally designed city is practised in the development of ideal cities and in the adjustment and improvement of existing cities, as best exemplified in the nineteenth-century transformation of Paris, which subsequently became crowned the capital of modernity (Harvey, 2005; Laugier, 1977). It is widely practised in the construction of colonial towns around the world, as part of the conquest of nature, which for the European colonists included the local populations. It is the powerful states and production capacities of the later centuries that finally enable the materialization of that vision of building completely new cities and redeveloping entire parts of existing cities. The practice of designing entire cities culminated in modernist urban redevelopment and in the garden city movement since the early twentieth century. Although the shape of the political power had changed from single rulers to complex bureaucracies, there was a continuity in uniting the political and aesthetic powers. The search for authenticity in single sources of authority is inevitably a political undertaking.

### *Complexity of city building*

This line of linking a city's quality with the authorship of a designer is close to the concept of autonomy and the rule of reason, but as we discussed earlier, autonomy and authenticity are not the same. Having a single designer for a city might make it more rational (although this is debatable as the rationality of a single person is not equated with the rationality of an entire city), but this rationality would not make the city more authentic. Authenticity in this case would reside in the realm of symbolism and reputation, whereby it is the status of the city, rather than its rationality, that becomes related to the identity of its designer. The city is treated as a work of art and its designer as the genius artist who produced it. This linkage is hoped to create a positive image for the city, asserting a status that becomes an asset in the global struggle for recognition. In addition to the rationality of the city, as Descartes had envisaged, it is also its symbolic value that is hoped to be enhanced through its association with a single designer. If there is a discussion about the authenticity of Chandigarh, it is directly related to the fact that it was designed by Le Corbusier. We can dispute whether this was a rational city, but we cannot deny that its global reputation was primarily made by its links with the world-famous architect. If there is a discussion about the authenticity of Brasilia, as UNESCO has acknowledged, it is through its connection to its designers Lucio Costa and Oscar Niemeyer.

In the context of neoliberal global competition between cities and states, this generates an added layer of significance. As Descartes noted, it is not often possible to redesign cities as a whole, but it is possible to design single buildings. Cities may not be completely redesigned, but their reputations can be remade through iconic buildings. The so-called starchitects are the authors of these buildings, which become the means of remaking city reputations in the context of global competition. Therefore, a city can claim to have an authentic Zaha Hadid or Richard Rogers building, which can be distinguished from those that were designed by the members of their office and be considered superior to those designed by other architects. This does not make the city authentic, but the authority of the celebrated designer provides some symbolic value for some buildings. This claim for status tends to reduce the complexity of a city to an image and claim the authenticity of that image, and by extension the authenticity of the city, by relying on the cult of genius.

Therefore, the rationality of a single designer, and the heightened status of that designer, are two different ways in which the designer is treated as an author of the city. In the first sense, the designer is seen as a scientist who can take different considerations into account and propose a rational order for the city. In the second sense, the designer is seen as an artist, a genius at work who can create masterpieces that would enhance the city's status and reputation. The authority of the single designer is established as both a scientist and an artist. In critical response, however, the agency of the designer comes under scrutiny. How far is the single designer responsible for the design, and what are the limits of a designer's power? Who else is involved in the making of the city, and how far can the design shape the city as a material and social environment? In later chapters, we will return to the questions about the city as a material and social environment, so in this chapter, we focus more on the question of power and agency of the author. As authenticity is claimed by virtue of tracing something back to its originator, the questions that need to be asked are about the originators and the conditions and limitations of their power.

The first question is whether genius designers invent the design through their own unchained intuition, or that they are operators within a cultural context. This is the question that we saw Barthes and Foucault raised about the role and significance of the author of a text. If designers use a public language of design, with its own concepts, grammar, tools of communication, and precedents, then they are not acting in a void but are bound by a series of possibilities and limitations. An early indicator of this mediation of a cultural apparatus is the way the invention of perspective created a new way of seeing, depicting, and designing the city (Iversen and Melville, 2010). The rules of perspective were not just an attempt to show the world as it is, but to establish a new way of seeing and an orderly depiction of that world. It was this ordered vision that was used in the design of ideal cities and the protocols of design that were subsequently deployed. Therefore, the designer operates with a set of common tools and draws on existing reference points, which shape and facilitate, but also limit, the design. The designer is not an omnipotent actor working in a void, but part of a network of actions and reactions, possibilities and limitations.

Moreover, much of design is a multidisciplinary group activity involving different designers, engineers, and other experts, so the proposal for a scheme, especially if they are at an urban scale, is the outcome of a series of deliberations and discussions within the group and with the outside stakeholders, rather than the private inspirations of a sole creator. The final design may carry the emblem of that sole creator, but it is often the product of a complex, multidisciplinary process of trial and error. The celebrated author may be no more than an editor, a team leader, or an office figurehead, rather than a genius acting alone. This is not questioning the talent and innovativeness of the designer, but to show that in major architectural, urban design, and city planning schemes, the design is a complex and multiagency process that draws on the protocols and precedents of a design culture and is located within social, economic, and political considerations (Madanipour, 1996, 1997).

A further question is about the materialization of the design after it leaves the office, about the complexity of producers and regulators and their relationships (Madanipour, 2006). Large projects involve many teams with diverse skills, each with an agency of its own, which can affect the scheme. The development industry, which includes the designer, produces a scheme, but on the way to implementation, it makes many adjustments that may not have been anticipated. In addition to the producers, there are also the regulators of the scheme who have their own inputs, both during the design process and afterwards. Rather than the simple link between a single ruler and a single designer, which would have been the case for Filarete, now the urban authority is often a complex bureaucratic organization which may be subject to democratic checks and balances. Even when the whim of an autocrat determines the fate of an urban project, it has to be coordinated through a complex urban and national bureaucracy, economic resources, and technological instruments and capacities, which will have their own impact on the process and its outcome.

The overall question is: who builds the city? An urban design scheme is handled by producers, regulators, and users, each with multiple impacts on how the scheme evolves and materializes (Madanipour, 2006). In such a multiplicity of actors, with their range of interests and powers, can there be a unity of authorship and purpose, or is the city the outcome of the interaction of many forces? Even if they all act within a zeitgeist, a single paradigm that can bring these people together, no paradigm is total, and no interest is paramount; there will always be diversity and multiplicity in intention, action, and conditions of possibility. The singularity of the author is a blanket concept for the process of production, the division of labour within it, and the tangible, physical, and social outcomes. But these outcomes are always pregnant with new horizons and can never be limited to this moment's conditions. Long after the designer and developer have left the scene, people who inhabit the area continue to make changes, as we will see in the discussion of time and temporality in Chapter 6 (Figure 3.5).

Cities are not authored by any single author. Even when a single person has been very influential, such as Baron Haussmann in Paris, Joseph Chamberlain in Birmingham, or Robert Moses in New York, or when a single designer has designed an entire city, such as Le Corbusier in Chandigarh, or

*Figure 3.5* Cities are continually built and rebuilt, by many hands and in a complex process (Toronto, Canada)

Lucio Costa in Brasilia, a city remains a collective and continuous creation, as it accumulates its characteristics through time and made by many. If there is any claim for authenticity, it would lie in the relationship between the city as a changing socio-spatial complexity and the myriad of authors that are involved in making it, including its residents who will make many changes through time. Cities do not have single authors, even if history attributes the construction of a city to a single powerful individual. It is the masses of workers who build the city and the people who live in it, who make and remake the city. If there is any point of reference for the authenticity of a city, it would lead to this complex set of agents, instruments, and relations, that are all involved in making and transforming the habitat.

## Authors and authenticity

Authenticity is a relational quality that shows a direct link between someone or something and an origin. One of the most important origins is the maker, the author, the artist, and the designer, whose identification as the source of a production would qualify that production as authentic. The author is presumed to be powerful enough to produce the self, the text, the work of art, and the city. Two major articulations of this powerful author are the notions of autonomy and authenticity, one drawing on intuitions to make a rational independent self in response to social restrictions, and the other drawing on self-discovery to construct a unique self in the face of uniformity and meaninglessness. The cult of the independent self is also present in the power ascribed to the author of a text, to the genius artist who creates a work of art, and to the designer who reorders the city through singular imaginary powers.

However, it is never possible to find an independent author who is detached from society and material environment. As the history of science, and the critique of the text and the work of art show, the author is never the sole origin of a work, as authorship always draws on existing structures, tools, and resources. Authorship is a work of mediation and engagement, rather than pure utterance. When it comes to cities, it is even more difficult to identify a single origin for a complex urban environment, either in the political ruler or in the designer. The authority of the author/designer as the origin of a self, a text, a work of art, or a city is always part of a much wider network of power relations. It is in its reception that the text, the artwork, and the urban space find meaning and value. Any sense of authenticity would lie as much in the interpretive experience as in the complexity of production and the intentions of the designers/authors. Authenticity becomes a relation of status and recognition, rather than rationality and functionality. The search should not be limited to a reductive dichotomy between self-determination and social determination, as authenticity is relational and dialectical, and investigating the interface between intentions and conditions of possibility, and the range of agencies and processes that are involved would provide a more productive ground.

The desire for freedom from social pressures has led to a demand for independence and self-determination, expressed in autonomy and authenticity, and in becoming the author of yourself and your environment. The problem, however, is that the author is never a singular source acting in a void, but an agent embedded in a series of relations, working with others, and drawing on different resources and tools. In other words, the search for the author does not lead to a single address, and the power of the author is far more limited than claimed; therefore, authenticity cannot be established through the singularity of the authorship.

# 4 Substance

Another way of evaluating the authenticity of something is about the object itself, its material, quantity, form, contents, composition, and methods of production. It is about the matter and substance as the origin, rather than the author, the artist, or the maker: the question is now about what it is made of, rather than who made it. One aspect of this approach pays attention to the originality of the material, and the processes by which it is made. Handmade objects find value, not necessarily because of a known maker, but due to the quality of the material, the individuality of skills that are involved in their making, as well as the relationship of the object to a particular place, time, and culture. This can also be found in the trend for healthy and environment-friendly food, whereby organically produced food from local sources may indicate a degree of authenticity. In a work of art, its uniqueness may be used as a mark of authenticity, distinguishing it from mass-produced objects. In urbanism, this focus on the object may be seen in the use of technology and materials, which has an impact on the relationship between the outside and inside, between the forms and contents of spaces. These relationships reveal shifts and tensions in the making of the urban environment, shedding light on the materiality of the economic, political, and cultural formations, rather than the wrongly assumed immateriality of the digital culture. This chapter examines the implications for authenticity of the shift from author to object, with its 'aura', uniqueness, quantity, materiality, form, and content.

## From author to object

In a famous 1935 essay, entitled The Work of Art in the Age of Mechanical Reproduction, Walter Benjamin (1969:218) finds the 'outmoded concepts, such as creativity and genius' to be in danger of being used by fascism. Therefore, he searches for a theory of art that would help formulate 'revolutionary demands in the politics of art', a theory that would be 'completely useless for the purposes of Fascism' (Benjamin, 1969:218). Fascism, as practised at the time in Germany, had turned politics into aesthetics, against which Benjamin suggests that aesthetics needs to be politicized. To do so, he shifts the focus from the artist to the artwork, looking for authenticity in the work itself,

DOI: 10.4324/9781003451778-4

rather than relying on the cult of genius and the authority of the author. He argues that 'The presence of the original is the prerequisite to the concept of authenticity' (p.220), but here originality lies in the unique history of the work rather than in the superiority of the originator. In this shift of focus, Benjamin pays attention to the production of the work, rather than the producer, analysing the impact of industrial methods of mass production on the originality of the artwork.

The shift from author to work, from subject to object, questions the authority of the author as the presumed origin of authenticity, as it could have provided a possible basis for authoritarian interpretations. Instead, this shift transfers the focus of significance in art from the artist to the artwork, which is a precursor to the later generations' transfer of significance from the author to the text and more broadly to the language. For Benjamin, this transfer is a political act, criticizing both the capitalist celebration of individualism and the fascist fascination with the cult of the leader.

This shift away from the author is also a central theme in the work of Benjamin's friend Theodor Adorno, whose theory of aesthetics also rejects Kantian aesthetics based on the notion of genius. As he writes, 'From the start, the genius aesthetic shifted emphasis toward the individual... and away from society by absolutizing this individual' (Adorno, 2002:170). Adorno is critical of the existentialist emphasis on authenticity, which he sees to be a recipe for detachment from society (Adorno, 1973). For him, 'The part played by the empirical I is not... the locus of authenticity' (Adorno, 2002:167). He does not reject the notion of authenticity, as evidenced by the frequency with which he approvingly refers to authenticity in art, but he looks for it in the work rather than in the artist. The shift from artist to artwork is not detaching the work from its context and seeing it as a pure aesthetic experience, but it is a shift from individual to society:

> The emancipation of the artwork from the artist is no *l'art pour l'art* delusion of grandeur but the simplest expression of the work's constitution as the expression of a social relation that bears in itself the law of its own reification: Only as things do artworks become the antithesis of the reified monstrosity. Correspondingly, and this is key to art, even out of so-called individual works it is a We that speaks and not an I.
>
> (Adorno, 2002:167)

Adorno does not deny the value of knowing about the artist, and in this, he concedes to Kant: 'One sees a painting differently if one knows the name of the painter. No art is presuppositionless' (Adorno, 2002:179). But for him, art is an expression of society and history, rather than an individual who stands above them:

> The historical moment is constitutive of artworks; authentic works are those that surrender themselves to the historical substance of their age

without reservation and without the presumption of being superior to it. They are the self-unconscious historiography of their epoch.

(Adorno, 2002:182)

This statement clearly describes Adorno's criterion for what makes an artwork authentic, which is to be unknowingly mirroring its society and age. However, this criterion, and the argument made on its basis, can be questioned, as no work is ever outside its society and age, and artists and artworks are not floating in a void but always part of a social and historical

*Figure 4.1* New information and construction technologies enable the development of unique buildings. Are these mechanically produced objects authentic, endowed with an aura, and reflecting their epoch? To whom and in what sense? (Los Angeles, United States)

context. Famed or nameless, innovative or mediocre, all artists are parts of their society, mirroring their world, knowingly or unknowingly, even when they try to go beyond by recounting the past or envisaging an alternative world. By this criterion alone, if interpreted narrowly, all artworks, and more broadly all objects, would be authentic, which makes the notion of authenticity meaningless (Figure 4.1).

Adorno's concept of authentic art, however, should be seen through the prism of his social critique. As a result, for him, an authentic work of art is an act of resistance against the culture industry, which commodifies the artworks and produces formulaic products, and against narcissism, which is a feature of the late capitalist society (Zuidervaart, 2007:17). As such, the authentic artwork is not seen as a mere reflection of its epoch, but a self-critique that reveals society's shortcomings. While every object may be taken to mirror its world, for Adorno the notion of authenticity embodies a critical social approach.

By moving the centre of gravity from artist to artwork, Adorno and Benjamin offer a political critique of liberal individualism, existentialist voluntarism, and fascist aestheticism. Descartes and Kant had argued for the power of the individual in rational self-determination, liberalism had advocated individual freedom of choice, existentialism had suggested the possibility of constructing an authentic life like an authentic work of art, and fascism had turned politics into an aesthetic performance. Adorno and Benjamin place the emphasis on the artwork as the product of a specific history, which enables them to interpret it critically through the prism of society, rather than the mysterious creation of a genius. It is based on the Hegelian notion that thoughts and actions are mediated, rather than contextless and detached. As Adorno (1984:167) puts it, 'according to Hegel's dictum, there is nothing between heaven and earth that is not mediated', and therefore Adorno rejects the notion of unmediated inspiration and action that is often attributed to the genius.

Suggesting that there is a 'We' that speaks through the work draws on the Hegelian notion of spirit that finds reflection in individual ideas and actions. The problem, however, is that the Hegelian notion of spirit can be interpreted in reductive ways, trying to attribute single characteristics to complex society and periods of history. But such a 'We' would always be plural and partial, as society is never homogenous and singular. As it has become common in recent years, it is always important to ask who this 'We' is who speaks. It may be used to avoid a sense of subjectivity, arrogant individuality, and personal authority, and to show humility, objectivity, and reliability. But it may also be a simple transfer from first person singular to first person plural, referring to an imaginary commonality that masks the partiality of views and interests.

The shift from the author to the work leads to a series of questions about the work. What is the work, what is it made of, and how can it find authenticity? What are the relationships between its materiality and its meaning, between its form and content? If not through its association with the author, how can its meaning be determined? Is there a need to find a clear meaning for such work at all? Does the work have intrinsic meaning, or does it need to be interpreted to find meaning, hence the significance of the viewer of the visual arts and, as Barthes argued, the reader of the text? Does the intention of the author matter at all, or is the meaning solely determined by the reader, viewer, audience, the citizen, in their encounter with the work? In cities, should meaning and authenticity be sought in the city itself, rather than in the intentions and actions of the politicians, planners, and designers? As there are many authors, viewers and readers, the human agency is not eliminated but multiplied. The locus of power is diffused by transferring it to the work, where it will be open for multiple and alternative interpretations. Is this a process of democratization or a loss of common frameworks and shared meanings?

## From object to experience

To find meaning, objects must be perceived and interpreted by someone. This takes place through a combination of immediate experience, mediated interpretation, and intersubjective recognition. When extended to a multiplicity of objects, interpreters, and social patterns, the challenges of finding meaning for objects and evaluating their authenticity become magnified.

Benjamin distinguishes authentic objects, which are unique and imbued with an 'aura', from mechanically reproduced objects, which do not possess any of these qualities. Aura is a key concept for Benjamin; it is 'the authority of the object' (Benjamin, 1969:215), drawn from its uniqueness and its authenticity. For him, the authenticity of an artwork is its 'most sensitive nucleus'; 'The authenticity of a thing is the essence of all that is transmissible from its beginning, ranging from its substantive duration to its testimony to the history which it has experienced' (p.215). When a work of art is reproduced, it loses 'its presence in time and space, its unique existence at the place where it happens to be' (p.214). This includes a history of the changes that it has gone through, both in physical and social terms, that is, the damages the object has suffered through time, its different owners, and their traditions. To be authentic, the presence of the original is a necessity, which is lost in the process of reproduction: 'The whole sphere of authenticity is outside technical—and, of course, not only technical—reproducibility' (p.214). What makes it authentic is its originality, uniqueness, and presence, and a historical testimony that becomes interrupted by mechanical reproduction:

> And what is really jeopardized when the historical testimony is affected is the authority of the object. One might subsume the eliminated element in the term "aura" and go on to say: that which withers in the age of mechanical reproduction is the aura of the work of art.
>
> (p.215)

The authority of the object, its aura, can be found in both historical and natural objects. This authority of the object is what Benjamin uses to reject the authority of the genius, but this authority is challenged when unique objects are replaced with mechanically reproduced ones. The aura can only be found in unique objects, and it is broken when the object is taken out of its shell and tradition, reproduced mechanically, and is multiplied in many copies. Benjamin sees this, which is carried par excellence by film (and in our time by digital technologies), as a shattering of tradition and associated with mass movements. He observes the loss of aura in the age of mechanical reproduction, but he does not turn it into nostalgia; instead, he welcomes the revolutionary potentials of the mechanical arts such as film. With the loss of authenticity and aura in mechanical reproduction, the politicization of art can become a possibility:

> From a photographic negative, for example, one can make any number of prints; to ask for the "authentic" print makes no sense. But the instant the criterion of authenticity ceases to be applicable to artistic production, the total function of art is reversed. Instead of being based on ritual, it begins to be based on another practice-politics
>
> (p.224).

In her introduction to *Illuminations*, the book of Benjamin's essays, Hannah Arendt (1969) compares Benjamin to a pearl diver, who searches for and brings up unique objects, looking for value in the objects themselves, rather than in reviving their traditions. As Arendt mentions, a central figure for Benjamin is the 'flaneur' to whom, while aimlessly meandering the city streets, things reveal their secrets. The flaneur is like the pearl diver, searching for objects that can trigger new thoughts. Arendt (1969:44) quotes Benjamin, 'The genuine picture may be old, but the genuine thought is new. It is of the present'. The implications of these statements by Benjamin and Arendt are that the meaning of the work is revealed to the person who comes across them, rather than in the work itself; it is the experience that becomes the standard of authenticity, which is expressed in the notion of aura. The diver and the flaneur become the interpreters of the object, and their experience is the criterion of meaning. The shift seems to be from object to experience. It is not anyone's experience, but the person who can appreciate the aura and recognize the value of the pearl.

As Benjamin observes, while art has long been reproducible in large numbers, mechanical methods produce and reproduce artworks in completely new ways. From the ancient production of coins, bronze objects, and terra cotta to the medieval woodcut graphic art, etching and engraving, and to the early modern technology of print, the methods of mechanical reproduction grew more powerful. With industrial capitalism, these technologies reached new proportions with the invention of lithography, photography, and film in the nineteenth century. Photography and film facilitated the generalization of images of everyday life, as print had done with writing before. Mechanically reproducing sound, image, and movement would radically change the nature of art, ending its autonomy and detaching it from its historical basis in cult and tradition. For Benjamin, this change has both negative and positive implications, which he mainly explores in the primary mechanical art, the film. Making films is exposed to capitalist methods of commercialization and diminishes any direct, unmediated contact with the work of art, but it has a 'revolutionary function', generating new possibilities for analysis, discovery, and representation (Benjamin, 1969:229).

A technique like a closeup can show the ordinary life of the city in completely new ways, and so he suggests that 'filmed behavior lends itself more readily to analysis' and, therefore, 'The camera introduces us to unconscious optics as does psychoanalysis to unconscious impulses' (Benjamin, 1969:229–230). The 'hidden details of familiar objects' are opened up by the camera's closeup, both extending 'our comprehension of the necessities which rule our lives' and assuring us of 'an immense and unexpected field of action' (p.229):

> Our taverns and our metropolitan streets, our offices and furnished rooms, our railroad stations and our factories appeared to have us locked up hopelessly. Then came the film and burst this prison-world asunder by the dynamite of the tenth of a second.
>
> (p.229)

Dismissing the mechanically reproduced artworks as inauthentic comes face to face with their power of analysis and exposure, creating an ambiguous position about the authenticity of the artwork. The capacities of the film that Benjamin notes have been magnified by the digital technologies of today, in which these technologies are often an essential method of art making. Even earlier technical methods had a noticeable impact on art making, as exemplified by the use of camera obscura in Renaissance painting, helping the artist with the accuracy of proportions and the perspective of the scene that they painted. Would the Renaissance works that used this technique be called inauthentic? In our time, the use of computers in writing, image making, music composition and performance, and design, is so widespread that some areas of artistic activity cannot be imagined without digital technologies. Some have argued that 'art making has entered an age of global reproducibility' (Zuidervaart, 2007:37). Finding the aura in such works becomes a serious challenge, but if the aura is a function of the viewer's experience, then aura can be found in any object, however produced.

Arendt asserts that Benjamin is closer to the existentialists and Heidegger than to his Marxist friends. As she writes, Benjamin 'replaces content with pure originality or authenticity, something that only French Existentialism established as a quality *per se* detached from all specific characteristics' (Arendt, 1969:44). Perhaps this is why Adorno, with his emphasis on content, is critical of Benjamin's concept of aura, a concept 'which Benjamin described at once nostalgically and critically', and which 'has become bad wherever it is instituted and simulated' (Adorno, 2002:45). Works that are mass-produced may seek to show a semblance of immediacy and individuality, 'as if in the administered world such still existed' (p.45). If it is 'conceived nondialectically the theory of aura lends itself to misuse', becoming 'a slogan of the deaestheticization of art' (p.45):

> The failure of Benjamin's grandly conceived theory of reproduction remains that its bipolar categories make it impossible to distinguish between a conception of art that is free of ideology to its core and the misuse of aesthetic rationality for mass exploitation and mass domination, a possibility he hardly touches upon.
>
> (p.56)

**Between presence, interpretation, and recognition**

For Adorno, what makes a work's aura is its content, rather than its presence: 'Aura is not only – as Benjamin claimed – the here and now of the artwork, it is whatever goes beyond its factual givenness, its content; one cannot abolish it and still want art' (Adorno, 2002:45). What makes a work an artwork is 'that objective meaning that surpasses subjective intention' (p.275). In addition to its content, the mastery in the use of techniques and the high level of artistic skills are also a part of a work's aura, 'the memory of the hand that, tenderly, almost caressingly, passed over the contours of the work and, by articulating them, also mollified them' (p.214).

Kant's aesthetic theory had asserted that the work of art should be beyond serving an interest (Kant, 1987). Adorno argues that 'Benjamin's definition of aura touched on this inner-aesthetic element, though it relegated it to a past stage and declared it invalid for the contemporary age of technical reproducibility' (Adorno, 2002:310–311). However, in comparison with authentic art, 'degraded, dishonored, and administered art is by no means without aura', which uses it for entertainment purposes (p.311). This is visible in every closeup in commercial films that mocks aura: 'Aura is gulped down along with the sensual stimuli; it is the uniform sauce that the culture industry pours over the whole of its manufacture' (p.311).

Rejecting the idealist modes of thinking that give primacy and authority to the subject, Adorno advocates a materialist response: the primacy of the object. For him the lines between the subject and object are not clear-cut, and their relationship is not an equal one: the subject tries to grasp the object, but something in the object remains beyond the reach of this grasp, so subjectivity does not capture the object in its entirety. Meanwhile, the subject is itself an object, and part of the meaning of subjectivity is to be an object; 'the I is an entity', so both subject and object can be considered as objects (Adorno, 2004:183). 'An object can be conceived only by a subject but always remains something other than the subject, whereas a subject by its very nature is from the outset an object as well' (p.183). The subject is always an object, but an object is not always a subject, hence the primacy of the object in an unequal relationship. As a result, the precedence of the object would refer to both subject and object in their differences; it 'means to make progressive qualitative distinctions between things which in themselves are indirect; it means a moment in dialectics' (p.184).

This does not mean the immediacy of objectivity or supremacy of consciousness; it requires critical and dialectical reflection. Adorno stresses that 'without a knowing subject nothing can be known about the object', but as the subject is also an object, 'The object's preponderance is solely attainable for subjective reflection, and for reflection on the subject' (Adorno, 2004:185). As 'Consciousness is a function of the living subject', subjectivity and objectivity are intertwined, and neither is possible without the mediation of the other (pp.185–186). This would help the enlightenment to go beyond its traditional understanding, debunking the myth of an all-powerful subject, realizing the limits of the human mind, and recognizing the rootedness of subjectivity in materiality (p.186). The primacy of the object does not assign primacy to a material world that is inaccessible to human understanding, as Kant had suggested with his notion of the thing-in-itself, but arguing for the materiality of human subjectivity and therefore recognizing its possibilities, as well as its tensions and limitations.

These two critical theorists, Benjamin and Adorno, both advocate a materialist departure from the subject towards the object, moving away from the artist and paying attention to the artwork. For them, this shift is a political act that can undermine the fascist, capitalist, and Stalinist emphasis on authority and tradition. They indicate two different approaches to the object and its authenticity. For one theorist, the prime focus is on the presence of the work, whereas for the other it is the content that matters. The presence can decontextualize a work and look for a meaning that stands on its own in its immediacy, but the content locates the work in a social and historical context and expects it to represent that context truthfully even if unknowingly. One sees an aura that emanates from inside the object, the other an aura that comes from the outside. The two approaches come close to taking different positions on the spectrum that range from self-determination to social determination, as we came across earlier. They both look critically and dialectically for an immanent meaning in the object, but they develop different articulations of that immanence.

Do different people experience an aura, or the same aura, when encountering an object? Where does the aura lie? A masterpiece of art may not make any sense to someone who is from another age or society, and they may treat it as a mere decoration or even destroy it as meaningless and obsolete. The aura would speak only to the connoisseur, who can recognize the aura, feeling the presence of an atmosphere or appreciate a work's content. But even different connoisseurs may have opposing ideas about a single work, one praising the aura and the other not seeing any aura in the work. Moreover, the interpretation of an object is not limited to the experts, and anyone can have a personal sense of the object. The object's meaning, therefore, is not detached and standalone, but it depends on the context, interpretation, and experience; it can be judged to be authentic or inauthentic when seen in the context of its web of relations and against the criteria that have been set in advance.

The notion of experience has been identified as a key concept in Adorno's thinking (Zuidervaart, 2007:95). This is an active, or 'emphatic experience' (*Erfahrung*) which is 'something the subject undergoes in relation to a particular object in its nonconceptualized particularity' (Zuidervaart, 2007:95). It is transformative, novel, and directly related to the object. As modern society has diminished the possibility of this experience, what is needed is 'philosophical experience', which combines this emphatic experience with rationality, bringing together the conceptual and the nonconceptual. However, philosophical experience is not available to everyone. In other words, here too, we need a connoisseur to appreciate and undergo this form of experience, resulting in criticism of elitism levelled against Adorno. His concept of emphatic experience, as a reverse of Heidegger's authenticity and its associated problems, is not put to intersubjective testing (Zuidervaart, 2007:98).

In response, Zuidervaart (2007:101) suggests public authentication of truth, which goes beyond individualized, detached or privileged experience but is intersubjective and democratic. The direct experience of an object or event may be judged on the basis of an emotional response, through which we feel that something has been revealed to us in its true form. It encompasses the experiences of both the artist and the audience, as it expects to have arisen out of true experience and to generate novel experience. An 'origin for authenticity' is then shown 'to be a vision or an experience' (Zuidervaart, 2007:43). An authentic work of art is expected 'to be imaginatively disclosive of the experience from which it arises, of the artwork's own internal demands, and of a public's need for worthwhile cultural presentations' (Zuidervaart, 2007:43).

However, it is not only the experience from which it arises but also the experience that it generates in the audience. But the audience is diverse and will have different responses. Moreover, the dual experience may easily be misleading, by accident or on purpose. We may be fooled by a con artist, who has a set of tools that seem to be effective in convincing others of the true value of something. Even if there are good reasons not to trust someone, we may ignore those reasons because we think that person is sincere, purely on the basis of their emotional effect on us, which can easily be deceptive. This suggests that the judgement of authenticity needs to combine emotional response with some knowledge and understanding, which is located in a context.

This knowledge and understanding may be presented by the connoisseur, but it is based on the standards of judgement developed by a group of experts, the institutional and cultural context in which they are located, and the paradigmatic values that have arisen in a society. An object finds its meaning in a context, and this context is multi-dimensional. An object found in a field by a farmer may just look old but may have no further meaning for the person who found it; it is a qualified archaeologist who can make sense of the object by locating it in a historical context, however tenuously. An object that is exhibited in an art gallery is considered by the organizers and visitors to be a work of art, but if the same thing is left on a pavement, it may just go unnoticed by the passersby as an ordinary object.

Mass culture is often portrayed as a passive phenomenon, and people are seen as a formless mass exposed to the manipulations of the dominating powers. But if we see the experience of events as active engagement, then everyone's experience is as valuable and significant as the other, and each experience can potentially be as valuable and significant for a person as any other. If a mass-produced object provides a deep sentiment in one person, can it be dismissed as poor taste and lack of cultural refinement, and therefore devoid of any authenticity? The object may not be authentic, but the sentiment of the viewer is, as it could be a genuine, heartfelt feeling of unity with the object, even if the object is judged to be culturally inferior. It is an authentic moment, if not an authentic object. The source of this authenticity is the person who experiences it, rather than the object, or the author of the object.

In his work of art in the age of mechanical reproduction, Benjamin refers to authenticity as the uniqueness of the artwork vs reproduced ones. But what if the mechanically reproduced image becomes a singular experience for a person? This is often the case in popular culture, where uniqueness is not accessible, and people use mass-produced objects. This democratizes the access to and use of objects, but this democratization renders them inauthentic, as they are mechanically mass-produced and distributed. At the same time, the experience of encountering these objects may be singular and authentic, as they may inspire and interest individuals, even when the object is mechanically mass-produced. We see this in music festivals, where a performance is made for thousands, if not hundreds of thousands of people. Seen from the outside, this is a mass event, devoid of authenticity, as the artist repeats what has been mass-produced before, and the individual members of the audience are lost in the crowd. But for each individual, it may be a novel experience, a singular event with an aura that may remain in their memories for the rest of their lives.

Benjamin and Adorno offer two visions of authenticity when applied to objects: objects are authentic when they stand on their own and emanate an aura, or objects are authentic when they represent a society and epoch. Both require a combination of experience and knowledge to judge the authenticity of the object. However, a single object can have different impressions and interpretations by different people, and some of those experiences might have a potentially strong sensual impact on the person, beyond their everyday habits. This is an alternative way of thinking about the inside–outside relations, in which the object's presence and appearance, its outside, is linked with an interpretation that the viewer finds meaningful, an inside. However, the diverse, singular interpretations are derived from social conventions, and they also lead to the emergence of new patterns and paradigms. This connection between experience and interpretation, when seen in its social context, becomes a way of linking an appearance and an idea, uniting them as a single phenomenon. The modernist emphasis on the relationship between form and function is a way of reconciling the outside and inside, but it reduces the meaning to a utilitarian one.

The shift from artist to artwork, therefore, would inevitably lead to a further shift from object to experience, either as grasping the aura of presence or comprehending the socially rooted critical content. The object carries an accumulated meaning that can only be unpacked and understood by the person who experiences it and knows about its location in a cultural context. The question, however, is whether this understanding is based on refined knowledge and skills, thereby privileging the educated and the expert, or based on immediate experience and common sense, which could democratize the sense of authenticity with or without any knowledge and skills for refined interpretations of the work. There will always be a diversity of interpretations about the authenticity of something, as determining it is not a scientific endeavour but an aesthetic judgement within a diversity of experiences and views. However, while experience is important, it is not sufficient for judging authenticity. If authenticity is establishing a truthful relation between inside and outside, then the experience alone cannot determine that truthfulness, as some knowledge of the object's features and contexts plays an important role in determining the viewer's aesthetic response. This indicates the socially mediated character of both the experience and the knowledge of the object, and thereby the social character of the authenticity, even when it is felt to be a detached, individual experience.

## Quantity and quality

Uniqueness has been widely used as a key criterion of authenticity. In search of a meaning for authentic urbanism, I decided to consult the nemesis of authenticity, that is, a machine. According to ChatGPT, an Artificial Intelligence program,

> An authentic city is one that maintains a unique identity and character, often shaped by its history, culture, architecture, and the daily lives of its residents. Unlike cities that may feel overly commercialized or homogenized, an authentic city preserves its distinctiveness and sense of place.

As the nature of the AI entails, its definition of an authentic city is a mechanically produced description that reproduces and echoes the current beliefs in the qualities of good cities. If AI existed at the height of modernism, it would probably have reproduced the maxims of modernist urbanism as the key features of an authentic city. As it stands, its definition of the authentic city feels like the vision of a city from the perspective of a visitor, a tourist, looking from outside in, searching for a unique character and experience. It is a skin-deep image of the city as it appears to the visitor, without seeing the complexities, hierarchies, and diversities that lie under the surface, or even visible to the sensitive observer. The AI definition stresses the uniqueness of identity and character, but the fact is that every city is unique, and every life in every city is unique.

Benjamin and Adorno discuss works of art, which have historically been unique objects before the arrival of mass production technologies. The uniqueness of the object would be an intrinsic part of its authenticity, but first it needs to be classified as a work of art. There have been long debates about what can be considered a work of art, but Benjamin and Adorno asked what makes a work of art an authentic one. In this discussion, the quantity (uniqueness) and quality (aesthetic value) of an object can potentially turn it into an authentic work of art. This raises at least three questions. First, can objects that are not works of art be considered authentic and be endowed with some aura, or is authenticity limited to artworks? Subsequently, if non-artworks can be considered authentic, what qualities do they need to be considered as such? Second, is authenticity a fixed and permanent feature of an object, or is it something that can evolve and change, or be judged differently by different observers? Third, can multiple quantities of objects be considered as authentic or is authenticity only a feature of singular objects?

The first question is about whether authenticity can be found in objects that are not artworks. If the works of architecture are considered singular artworks, then we could expect that the assessment of uniqueness and authenticity might be applicable to them. But we know that most buildings are not seen as artworks, as the architectural historian Nicholas Pevsner famously distinguished architecture from buildings. For him, the 'aesthetic appeal' was the criterion for making the distinction. 'A bicycle shed is a building; Lincoln Cathedral is a piece of architecture' (Pevsner, 1963:15). He explains: 'Nearly everything that encloses space on a scale sufficient for a human being to move in is a building; the term architecture applies only to buildings designed with a view to aesthetic appeal' (Pevsner, 1963:15). It is this aesthetic dimension that turns a building into architecture and thereby becomes a work of art. It is this aesthetic dimension that is meant to be captured in the concept of aura, as both Benjamin and Adorno discuss it in their writings on art and aesthetic theory. Does this mean that the bulk of a city, which is a collection of buildings rather than works of art, can never be considered as authentic?

After all, any city is a unique gathering of lifeforms and inanimate objects in a unique location. No two cities are alike, and therefore if uniqueness is the primary criterion of authenticity, then all cities are authentic. Even if two cities are designed to be exactly the same, the course of life in the city will soon diversify them into two distinctive cities. Many cities are criticized to be 'cloned', filled with the same global brands and their standardized designs. But calling these cities as cloned ones does not mean that they are the same, as each is unique by virtue of the particularity of its contents and composition.

What Benjamin and Adorno both discuss are the criteria for the authenticity of artworks, and both consider that criterion to be having an origin, albeit they define this origin differently: for Benjamin, it is a biography that is reflected in a presence, and for Adorno it is a social and historical context that is reflected in an expression. This tracing of an origin means that non-art objects could also have the possibility of authenticity if they are shown to be related to an origin. But as we see in the work of only two theorists, there is no common view about that origin. What is that origin and where can it be found?

The second question asks whether authenticity is a fixed and permanent feature. If authenticity is a quality of the relationship between an object and an origin, it can only be fixed if both the object and origin remain unchanged and our views of that relationship never alter. However, as all these three elements are subject to change, authenticity of a thing cannot be fixed and permanent. The object goes through a process of transformation, as all material objects are exposed to decay and loss of energy or are repaired and reshaped. The origin is a moveable point of reference, which is exposed to the difference of opinion and preference. An object that is imbued with an aura for one group is utterly meaningless for another. The relationship between the two, which is a process of assessment and judgement, may be performed by different groups and with differing outcomes. The result is that authenticity is a context-specific and observer-dependent value judgement, and hence it cannot be a fixed and permanent feature of something.

The third question is about the quantity of the objects. Benjamin and Adorno discuss the works of art, which historically have been unique, but can we find authenticity in multiple objects? Benjamin's argument is that authenticity disappears in mechanically reproduced works. According to this view, in the industrial age, most things are mechanically (re)produced, and so authenticity becomes a thing of the past. However, it can be argued that everything is unique, even when produced through industrial mass production. It is true that handcraft objects have a larger number of unique features, but mass-produced objects are not identical, each has a different materiality and existence, and each is experienced differently, going through a journey that locates it in a different context, carrying meaning or even emanating some aura for someone. Everything is a singularity, no matter how it has come into being. By this token, everything has the potential for being valued as authentic.

The distinction between craftsmanship and mechanical reproduction is a fundamental topic in the analysis of authenticity. An artefact made by a craftsperson is unique, displaying elements of its maker's individuality, even if that person makes many copies of the same object. In contrast, a mechanically produced object does not show the individuality of a personal author, and it is often taken to be an impersonal, soulless object. The nineteenth-century arts and crafts movement was a reaction to the industrial production methods, which were seen to have drained the individuality of objects, turning them into mass-produced, utilitarian devices. The critique that Camillo Sitte (1986) raised against the modern reorganization of Vienna, in which the old walls were demolished to pave the way for wide boulevards and monumental buildings, was a concern for the loss of artistic sensibilities in city design. To him, the new urban projects compared poorly with the medieval cities, which, with his arts and crafts approach, he found much more aesthetically appealing.

From our vantage point in time, however, some of those mass-produced objects of the nineteenth century have become historical objects with a level of individuality and authenticity that may not be found in later objects. An example is the continuing high demand in the UK for Victorian houses, which are sought after by many as having 'character'. These houses may have been produced by the volume builders in styles that are replicated across the country, and primarily in pursuit of profit by its developers and in response to the growth and suburbanization of cities. But after two centuries, these mass-produced buildings have found a sense of uniqueness that distinguishes them from later housing styles. The complexity of technology used in these buildings may have been much simpler than the technologies of today, but they seem to appeal to our contemporaries as something desirable due to their differences and their accumulation of prestige over time.

By our time, as the mid-twentieth-century modernist buildings are threatened with demolition, many attempts are made to safeguard them as cultural symbols of their time. This would give them a sense of authenticity, an aura that reflects a culture at a specific period, even if these buildings were designed and made according to industrial methods of thinking and constructing. What these two examples of the appeal of the Victorian house to the twenty-first-century urban households and the appeal of the modernist buildings to the cultural connoisseurs of our time show is that the concept of aura and authenticity are partly based on the image we construct of the past, without much attention to the ways in which things are produced.

Moreover, the criterion of quantity is open for questioning. Every object is made of other objects, a combination of the primary elements that constitute the world, a composition of particles that are also found in other objects. As such, nothing is purely unique, but always a constellation of other elements. According to this argument, nothing is singular, and we need to go down to subatomic levels to reach a degree of singularity, if at all. When it comes to cities, which are made of many animate beings and inanimate objects, cities are as much a singularity as a single building. A single building is a collection of many elements and spaces, and so is a city, albeit at a larger scale and degree of complexity. Furthermore, a city can generate a singular experience, be imbued with a distinctive atmosphere and character, which can bring all these disparate elements into a single perception. A city can become as singular as any other object: it is similarly made of many elements, and it can convey a single impression, and have a distinctive aura, which can only result from its particular composition of elements, such as people, houses, and streets, that may also be found in other cities but in other forms and compositions. The criterion of quantity, therefore, loses its potency in determining authenticity.

It is therefore difficult to establish the authenticity of something by limiting the notions of quantity and quality to singularity and aesthetic content. Multiple, non-artistic objects can also be found to have authenticity, and similarly, the authenticity of singular artworks can be questioned. This does not mean a relativization of authenticity, or simply denying its existence, but realizing that determining authenticity is a process of judgement that cannot be undertaken in a void, beyond the conditions and contexts in which the judgement takes place.

## Matter

Authenticity may refer to the material with which an object has been made. If a furniture maker claims to offer authentic wood furniture, this claim sets wood as the basis on which the authenticity of the furniture can be established. It means that this table is made of solid wood, rather than engineered wood, that is, chipboards made of sawdust and small wood chippings glued together by resin and enveloped in layers that give it the appearance of solid wood. The same claim may be made by jewellers comparing gold and gold-plated items, or between natural and cultured pearls, which distinguishes two types of origin from each other, one found in the natural world and the other in manufacturing processes. In other words, these claims to authenticity make a distinction between non-human and human-made materials, making a value judgement in which the latter is found to be inferior to the former. Wood, gold, and pearl that are produced through the growth of trees, formation of the earth, and defensive mechanisms of oysters and mussels, are given a higher priority over the compressed sawdust, electrochemical layering, and lab-grown gems that are manufactured and thereby capable of being mass-produced. One group is seen as natural and original, in contrast to the other, which is fake and artificial.

This distinction is about quality rather than quantity. In this context, to be qualified as authentic, objects do not need to be unique, but to be made of an original material. There are many wooden tables, many golden rings, and many pearl necklaces, and they can all be claimed to be authentic in terms of their main material. Authenticity describes the relationship between an object and its origin. But if not uniqueness, what is the criterion for selecting these origins? What is the level of detail that we need to zoom into to find that origin? Both natural wood and chipboard are made of wood, so how do we distinguish between these two wooden origins? Is the distinction based on the processes that make that wood: slow growth of a tree vs fast speeds of engineering? Or is it based on the way wood is prepared for use in furniture and in buildings: cutting and shaping a versatile material vs creating the material to the required size? How do we select the constituent material? Do we go down to the chemical composition of these origins, to find deeper origins? Wood is primarily made of carbon and oxygen, but we do not determine the authenticity of the wooden furniture by the amount of carbon.

This means that the selection of what counts as origin is at the level of common conventions and understanding, rather than further detailed levels of physical and chemical composition. In the case of these examples, and indeed all objects that are evaluated for their authenticity, the origin is selected from a number of possibilities, and it is given priority through some socially and historically determined criteria. Therefore, the authenticity of materiality is evaluated at a conventional and everyday level, that is, as a social concept rather than a scientific analysis of its constituent elements.

The social concept of an object may include a number of characteristics. Natural wood's features that give it priority over manufactured wood include durability, versatility, strength, and aesthetic diversity. Another feature is its extent of exclusiveness and rarity. As manufactured objects can be made in abundance, rarity becomes a value. Gold, precious gems, and natural pearls are limited in their numbers, in contrast to the manufactured items of jewellery that can potentially be made without any limitation. This takes us back to Benjamin's discussion of mechanically reproduced objects, and while for him it was the artworks that needed to be unique, we find that in other objects rarity is also a source of value. Rarity is a quality that is based on quantity, and it is perhaps one step below uniqueness in quantitative terms.

Placing premium value on uniqueness or rarity of objects is an old social habit at the heart of economic reasoning. It is used to describe the relationship between supply and demand, whereby the reduction in supply would make an object less accessible, and if the demand is not reduced, the value of the object rises. Unique and rare objects find higher value if they are in demand and if this demand is high. Markets are partly based on this supply–demand relationship, and the economic value of most objects is determined through this calculation. While authenticity is not a quantitative value, it seems to follow a similar model, whereby value is assigned to an object through its level of availability and accessibility. Uniqueness and rarity suggest exclusiveness, beyond the reach of those who do not have the necessary social and economic resources.

However, it might be argued that wood is not such a rare material, and while natural wood can be produced through mechanically enabled forestry processes, it is still rarer than manufactured wood. But it is possible to produce pure wooden furniture in large numbers and through mechanical manufacturing processes. We may still judge that this table is a piece of authentic wood furniture without a view to the rarity of wood or the way that wood was cultivated but to the accuracy of description. An object may be judged to be authentic in one aspect but not others: the table is authentic wood furniture, but it was made yesterday in a standard design, so it is not authentic in its design or historic value. Calling it an authentic piece of wood furniture establishes a direct link between furniture and wood, between an object and a material origin.

In architecture and urbanism, this relationship between object and origin became an important topic of concern in the modernist movement. Historically, the material that covered the façade of a building could be different from what the building was made of. The widespread use of stucco made it possible to hide the materiality of the building behind an ornamental façade. For the modernists, however, adding ornamentation was a form of fakery; what was needed was to show the material in its naked form, so that the building would be an honest construction. It was argued that a building made of concrete should not hide this material behind ornamental details, colours, and other materials. Its structural properties and its façade should coincide. The building's inside and outside, what made it as an object and how it was presented to others, were to be unified and harmonized. Concrete may have been seen by some as brutal and ugly, hence modernist concrete buildings in the UK were given the term brutalism, while others praised its aesthetic qualities (Calder, 2016). The following period of postmodernism returned to praise of ornamentation, abandoning the quest for sincerity and authenticity in materiality (Jencks, 1992). The return of concrete to popularity in the 1990s was no longer based on a sense of authenticity, but for its plasticity and a transformed aesthetic sensibility, in the age of digital technologies, globalization, and iconic architecture.

Therefore, materiality can become a qualifying term for the authenticity of an object. It is an authentic wooden furniture, an authentic pearl necklace, and an authentic concrete building. But concrete is a manufactured building material, made of cement and sand, and as such would not be as natural as wood or gold. In any case, a building is so complex that it can never be built in a single material, and there will always be a series of different materials used for different parts and purposes of a building. To name a building as a concrete building means ignoring all the other elements that have been at work, giving priority to a single material. In this way, if the identity of an object is accurately described, on the basis of some material feature that is taken to constitute its originality, it can be described as a materially authentic object. This description may combine particularity and generality: all buildings that are built in concrete are in the same general category, and all may be called authentic concrete buildings, if concrete is seen as their primary feature. The claim to the authenticity of a concrete building would be strengthened if the building showed this material both inside and outside, as many modernist buildings did, rather than trying to hide behind ornamentation and colour.

However, it is not easy to extend this description from buildings to cities, which are made of many buildings with different materials and differing levels of correspondence between their appearances and their construction materials. The diversity in materials, and in the relationship between construction materials and facades, would make it almost impossible to find such direct relationship between a city and a material. Some historic towns and villages may be entirely built of a local material, but this may evolve and change over time, and even these places are widely diverse in their construction materials and appearance. A pearl necklace might not change much through time, but a city is a living organism, a diverse and evolving entity, continuously rebuilding and changing itself. It is possible to find a village largely made of stone, or a town mainly made of brick, but even here elements of change are visible, and the singularity of the construction material, if ever this was the case, may be disappearing. If authenticity is a description of an identity that links an object to a single material origin, most contemporary cities may not qualify as authentic in terms of their materiality. But as descriptions are often selective, it is always possible to reduce the narrative identity of a place to

*Figure 4.2* The city is nicknamed the Granite City for its widespread use of the stone (Aberdeen, UK)

its primary material, facilitating the claim to have, for example, an authentic concrete building and an authentic stone-built city (Figure 4.2).

## Form

Rather than the materiality of an object, its form may be selected as the basis for evaluating its authenticity. In this case, an object may be judged to be authentic if its form is original. Form is an integral part of the object, but selecting it as the origin of authenticity would suggest a shift of focus from matter to form. In our line of investigation into the relationship between the inside and outside, the shift from matter to form indicates a movement from the inside to the outside of an object, from what it is made of to how it appears to us. As an example, in a city we may focus on the density with which the buildings are grouped together or the pattern of streets; in this case, the form becomes the basis of any value judgement, rather than the building material. The form of the city is an element of its materiality, but it is different from its matter.

The relationship between form and matter was a fundamental preoccupation in ancient Greek thought, although their concepts of form and matter are more complex than the way these terms are used in the design discourses. As dramatized in a famous allegory of the cave, Plato (1997, *Republic*, BookVII) imagines that we live in an underground prison, and what we see is a play of shadows, but if we stand up and look towards the light, we see things in their true essence, where reality lies at a higher, more abstract level of ideas. Plato divides the world into sensible and intelligible parts. We live in the physical world, which we can grasp by our senses. But this world of multiplicity and change is just a mere surface, and reality lies elsewhere in a world of eternal, unchangeable Ideas or Forms, which can only be grasped in the light of intellect. Every object has a form, through which we can recognize what it is: it is through the form of a human that we can say this is a human, through the form of a city that we can call it a city. The form of objects that we see is an imperfect version of the perfect Form that exists at a higher plane as a nonphysical, universal, and eternal entity. True knowledge is to grasp this essence of things through abstract thinking.

In this Platonic system, the authenticity of objects may be difficult to find. If objects are imperfect copies of a perfect idea, and appearances are fleeting and faulty references to a permanent essence, then the possibility of uniting the sensible and intelligible worlds may not arise at all. These are two separate worlds and the best we can do is to recognize this distinction and try to reach beyond the senses and into the abstract realm of eternal knowledge. The form of objects is always a reference to their true core, the outside is always a replica of the inside, however imperfect and fragile. However, Plato and Socrates do discuss the gap between the inside and outside, which is where the topic of authenticity may be located. As we have seen in Chapter 2, Socrates approvingly mentions the maxim Be True to Yourself, but this recipe

*Figure 4.3* Is the authentic identity of a place defined by its materials, form, or content? (Vienna, Austria)

for a form of authenticity comes at a practical, rather than epistemic, level of discourse. An authentic object becomes one in which matter closely reflects the form.

In our analysis of materiality as the origin of authenticity, this separation of form from matter brings about some questions. Do we judge authenticity through something's matter or its form? In the case of a wooden chair, do we say this is an authentic piece of wooden furniture, thereby giving priority to its material, or do we say this is an authentic chair, thereby giving priority to its form? This takes us back to the moment of selection: what do we choose as the criterion for authenticity? It can be matter or form, and each will give us a different answer. But together, they show that the question revolves around the choice of origin, rather than a primordial element in the object. It is that choice that determines the basis on which the judgement takes place (Figure 4.3).

Rejecting Plato's separation of form and matter, Aristotle combines them in the concept of hylomorphism. For Aristotle, physical objects are compounds of matter and form, but what he means by matter and form has been the subject of debate (Ainsworth, 2020). He develops his concept in physics, trying to understand how different substances come into existence, and then extends this concept to other areas. For him, matter refers to all the component parts of an object, and form is what unites them into a single object. In this analysis, matter is the underlying nature of something, which is distinguishable from its form or substance:

> The underlying nature can be known by analogy. For as the bronze is to the statue, the wood to the bed, or the matter and the formless before receiving form to any thing which has form, so is the underlying nature to substance, i.e. the 'this' or existent.
>
> (Aristotle, 1991, *Physics*, Book I, 191a9–191a12)

Accordingly, bronze and wood are the matter and the nature of these objects, while statue and bed are their forms. But form is not limited to the shape of things; it is their definition or essence, which is closely related to their function: a human-shaped statue is not a human, as it does not function like one (Ainsworth, 2020). When form and matter are combined in every object and related to its function, this definition becomes more complex. For Aristotle, the nature of an object can be understood in four 'causes' or explanations: what it is made of (matter), what it is (form or essence), what brought it to existence (efficient or moving cause), and what its purpose is (final cause) (Aristotle, 1991, *Physics*, Book II). Therefore, in a house, its matter is wood or brick, its form is being a house, its making is a construction process by builders bringing together the necessary materials and skills, and its purpose is providing shelter. Depending on what is selected as the primary aspect of something, it would then serve as the basis for its evaluation. But as we see in this Aristotelian analysis, form, matter, agency/process, and function are all intertwined.

In Kant's aesthetics, the notion of form is closer to the design discourse. He separates the world into two parts: the world as it is and the world as it appears to us, splitting noumena from phenomena, arguing that we only have access to appearances, to forms. In Kant's critical theory, form is given prominence, as it reflects how our cognitive powers, with their own structure and limits, understand the world. 'A judgment of taste ...is merely *contemplative*, i.e., it is a judgment that is indifferent to the existence of the object' (Kant, 1987:51, I,I,§5). Aesthetic judgement is based on pure form, which should be free from any interest, but this form, especially in nature, has an inherent purpose. In other words, aesthetic judgement is based on 'a merely formal purposiveness, i.e., a purposiveness without a purpose' (73, I,I,§15). This purposefulness is not objective or functional, but intrinsic, which is why we expect others to agree with us in our judgement. It is 'connected with the feeling of a pleasure, a pleasure that the judgment of taste at the same time declares to be valid for everyone' (Kant, 1987:66, I,I,§11). This emphasis on form is exemplified in the significance of design in the visual arts and architecture:

> In painting, in sculpture, indeed in all the visual arts, including architecture and horticulture insofar as they are fine arts, *design* is what is

> essential; in design the basis for any involvement of taste is not what gratifies us in sensation, but merely what we like because of its form.
>
> (p.71, I,I, §14)

The design does not include the colours because the 'colors that illuminate the outline belong to charm' and 'A *pure judgment of taste* is one that is not influenced by charm and emotion' (p.69, I,I, §13). In the second place after the design, it is the composition that is the proper object for a pure aesthetic judgement (p.72, I,I, §14). But how is it possible to judge something only on its form, as if the form was independent without a history and presuppositions?

Chapter 3 discussed the distinction between autonomy, as championed by Kant, and authenticity, as championed by Rousseau and later by the existentialists. While they both privilege the author, their distinctions are also clear. This distinction finds some parallels in the early twentieth-century art between expressionism and formalism. Expressionism was an attempt at reflecting the deep feelings of the artist, which would align with the phenomenological and existentialist movements. In representing the world as it was felt, expressionist art would strive to unify the inside and the outside by privileging the inside. In contrast, formalism focuses on the appearance of the work, the shapes and composition of lines, surfaces, colours, and textures, without necessarily delving into the content of the work. It stresses 'the primacy of the visual in the experience of art' (Iverson and Melville, 2010:60–61). It places emphasis on the outside and the visible, which define the work. Formalism has been criticized for its neglect of the content and the context, but it has played a strong role in the development of modern art.

Form cannot be interpreted as an entity that is completely detached from the world. Here, we have seen three interpretations of the form: those that locate the source of meaning in a higher order, those that locate it within subjectivity, and those that locate it in the form itself, even though Kant links the form with subjectivity and some intrinsic universal laws. Another possibility is to locate the meaning of the form in its content, which recognizes the social character of the object.

## Content

Beyond form and matter, an object could also be evaluated through its content. As discussed earlier in this chapter, the relationship between form, matter, and content has been interpreted in different, and at times oppositional, ways. Hannah Arendt (1969:44), for example, suggested that Walter Benjamin 'replaces content with pure originality or authenticity', which appears to indicate that authenticity and content cannot be reconciled, as authenticity is 'a quality *per se* detached from all specific characteristics'. In contrast, Adorno sees content as a necessary condition of authenticity in artworks (Adorno, 2002:45). For him, authentic works of art are 'the self-unconscious historiography of their epoch' (Adorno, 2002:182). In this argument, he draws on Hegel, who believes that things do not have a meaning on their own and their meaning is 'mediated' (Adorno, 1984:167). In other words, an object is always interpreted from a human perspective, so it can never be free of associations, and those associations are parts of a social and cultural context.

Aristotle's model of four causes avoids the competition between form, matter, maker, and purpose, as they are all seen as integral parts of an object (Aristotle, 1991, *Physics*, Book ii). In this model, content may be related to all four dimensions, but particularly to its matter and purpose. The four dimensions may be related in a variety of ways. For example, two things that are made in the same form, process, and purpose may be made of different materials. Moreover, it is not always possible to determine the purpose of things, especially in the natural world. What can be said about the purpose of a bird or a tree, on indeed a human? But in human-made world, most objects are made for a purpose, although they too can be used for other purposes.

In architecture and urbanism, this Aristotelian model translates into design, construction materials, production and management process, and function. Buildings, streets, and social institutions can be analysed through these four dimensions. But here too, the four dimensions are not necessarily connected. A building with the same material, production process, and purpose may be designed in many ways. And here too, the problem of determining the purpose of space needs to be addressed. A building may be designed and built for a particular purpose, like a house, school, or hospital, but it may be used for other purposes, as in many old buildings that go through conversion and reuse. Moreover, what is the purpose of a city? Can we identify a single, or a few purposes, for a city in all its diversity and multiplicity?

Modernist architecture redefines the four Aristotelian elements in its own way, without any reference to the ancient philosopher. For Le Corbusier (1986), the form should be pure geometrical shapes, which are commonly used by the engineer, but not so by the architect. 'Working by calculation, engineers employ geometrical forms, satisfying our eyes by their geometry and our understanding by their mathematics; their work is on the direct line of good art' (Le Corbusier, 1986:2). Materials should be the modern industrial materials of concrete and steel, which 'in the last fifty years ... have brought new conquests' (p.7). The process and method of construction should be mass production of houses as machines: 'Industry on the grand scale must occupy itself with building and establish the elements of the house on a mass-production basis', while 'Economic law inevitably governs our acts and our thoughts' (p.6). The purpose is to solve a major social problem: 'The problem of the house is a problem of the epoch' (p.6), as 'It is a question of building which is at the root of the social unrest of to-day: architecture or revolution' (p.8). If the house is envisaged as 'a machine for living in' (p.4) and produced through industrial methods of mass production, a solution could be found.

In this design theory, content is given explicit priority and articulated as a plan and as a function. As Le Corbusier argues, 'The Plan is the generator'; the plan is not mere form, but it is the order of things: without it 'you have lack of order, and wilfulness' (Le Corbusier, 1986:2). Changing the plan means changing the order of life, which is badly needed in the modern city. 'Modern life demands, and is waiting for, a new kind of plan, both for the house and for the city' (p.3). This reordering is done through a rationalist, functionalist analysis.

The rationalist order is given a specific shape: the straight line, as it is through geometry that 'we perceive the external world and express the world within us'; the straight lines of pure geometrical forms are 'the material basis on which we build those symbols which represent to us perfection and the divine' (Le Corbusier, 1987:xxi). Le Corbusier famously distinguished humans from pack donkeys via a geometrical criterion. 'Man walks in a straight line because he has a goal', but 'The pack-donkey meanders along, meditates a little in his scatter-brained and distracted fashion' (p.5). In sharp contrast to the aimless ways of the flaneur, Le Corbusier's ideal human is goal-oriented and rational, considering the city as 'the grip of man upon nature. It is a human operation directed against nature, a human organism both for protection and for work' (p.xxi).

The subject of content and its relations with form is explicitly treated through function. The maxim by the architect Louis Sullivan (1896:408) became a motto of modernism: '*form ever follows function*, and this is the law'. Functionalism becomes fully expressed in the Charter of Athens, the modernist manifesto for urbanism, as drafted by the CIAM in 1933 (Sert, 1944:246–249). The city is to be reorganized through a new functional order, whereby four primary functions are each given appropriate spaces through a system of zoning. Housing is the primary function, but it suffers from high densities, overcrowding, and bad locations, whereas residential areas should be constructed with modern technologies on the best sites, with good exposure to sun, and away from the major thoroughfares. The second function is recreation, but open spaces are insufficient and poorly situated. The proposed solution is to raze slums in densely built neighbourhoods and use the space for recreation. The third function is work, but the industrial and business areas are not properly located, and there needs to be classification, zoning, and coordination between workplaces and residential areas so that travel and traffic can be minimized. The fourth function is transportation, but the urban streets are outdated for the machine age, so the solution is to create a new street network that is fit for motorized movement, whereby streets are classified according to their functions (Sert, 1944).

These principles of modernist architecture and urbanism show a fairly coherent design theory, which has been powerful enough to (re)shape much of the world in the last century. Rather than being 'the self-unconscious historiography of their epoch' (Adorno, 2002:182), it has been a tool for self-conscious reshaping of its epoch's built environment. Can this design theory be considered an authentic one on the basis of its coherence and reference? In its own way, it articulates, albeit naively, the problems and sensibilities of the urban industrial age. In its destructive treatment of the natural world, it expresses the dominant thinking of its time, at least in the centres of power. Can it be called an authentic voice of its era, by virtue of its extraordinary success in transforming the cities of the world? But doing so would reduce an era to a simplified formula. As the critical responses to modernist developments started to emerge from the 1960s onwards, it is possible to see that an entire era cannot be articulated in a single way (Gans, 1968; Jacobs, 1961). Nevertheless, if the coherence of a theory and its articulation of a period's spirit are the criteria for authenticity, that is, if a link between performance and period is established, this design theory could be seen as one of the authentic voices of its time. It advocated mass production, and so it would not qualify for Benjamin's test of mechanical reproduction, but the modernist high towers, as represented by Chicago and New York, emanated an aura that captured the imagination of the twentieth century.

By the same token, however, can we call the new metropolitan iconic towers around the world as the authentic representation of the age of globalized neoliberalism? (Figure 4.4). After all, they are built by and for urban authorities and global corporations as instruments of global competition and status assertion, tightly aligned with the era of global neoliberalism. As discussed

*Figure 4.4* Is the city's skyline an authentic representation of the age of globalized neoliberalism? (London, UK)

earlier, in this way any object can be considered an authentic object, as it cannot be outside its social and historical context. Thereby any distinctive meaning for authenticity would evaporate. The question becomes how far an object can find symbolic value to become the representative of age, despite all the doubts and shortcomings, and how that symbolic value is evaluated.

## Use

The works of art are not judged by their use, and indeed they have been characterized by their 'uselessness', that is, they are not bound to the calculations of utility. In design, however, functionality and use are major considerations. Both art and design involve creative exploration, but one is free from utilitarian necessity, while the other is an applied art, bound to the making of useful objects and places. The relationship between art and design is a relationship between ideas and practices, between expression and application.

When it comes to the urban environment, use becomes a major factor in the evaluation. Architecture is a public art, as its spaces are used by people and therefore need to meet specified expectations of functionality. Moreover, it is not sufficient that space be designed for a specific function; it also needs to be seen to be well-used. There is a clear distinction between function and use, as the design may predict and plan for a certain function, but the space may not be used in that way, or its use may change over time. Therefore, in addition to form and content, the authenticity of an urban area may be judged through its appropriateness and level of use. Design projects are presented, and judged, by how well they are used, which is partly an indication of their contents and functionality. There are, however, many other factors at work to make a well-designed place also a well-used one, factors such as location, reputation, accessibility, relationship to other activities, etc. It is not a given that good design would inevitably lead to the intended or extensive use.

Streets and public spaces are often taken to represent the city, as famously put by Jane Jacobs (1961). Busy streets are portrayed in urban design projects as indicators of the predicted success of a scheme. In other words, the image of authentic urbanism seems to be that of a busy street or public space, with happy crowds on a sunny day. Heavy use of space becomes an indicator of authentic urbanism. The meaning of authenticity in urban development may be taken to indicate that true urbanity requires social vitality, and therefore being true to itself for a newly built area would mean becoming a vibrant place. This can be seen in an example.

A new urban development scheme by the Cambridge University in Eddington, two miles northwest of the city, has aimed at providing 'modern, affordable homes for University staff – as well as all the services and facilities a community needs'; 'designed and built to be a neighbourhood – not just a place to live' (Cambridge University, 2023). It boasts to include a range of accommodation types, plus a primary school, nursery, shops, supermarket, sports facilities, open spaces, and a centre for community events and performing arts. Nevertheless, it was not judged to be successful, despite the need for accommodation in the city, proximity to the university, and using new methods to provide centralized recycling points and ultra-low water usage. The flats were not sold, the retail and hospitality lots remained empty, and the area's small squares and public spaces were 'eerily quiet' (Foster, 2023). According to Stephen Kelly, director of planning for both South Cambridgeshire district council and Cambridge City Council, 'It just takes time to create authenticity and for the market to absorb new houses' (Quoted in Foster, 2023).

*Figure 4.5* If vibrancy and intensity of use are taken as a sign of authentic urbanism, can a quiet street be also an authentic place? (Fez, Morocco)

This statement is based on the idea that authenticity in urbanism is reflected in its use, facilitated by the market, an authenticity that can be created by the planning and development process, but it takes time. What is implied in these statements by the city planner and the University is that a combination of residential units and associated services creates a community and that a thriving area is the indicator of authentic urbanity. If the area is full of people and activities, it becomes authentic: residential units are sold, and the relevant services and facilities are eventually up and running. In other words, it becomes authentic when a thriving urban area has been created. The question that emerges, however, is whether every busy urban area that includes these different functions can be called an authentic one. Does authentic urbanism mean having a busy urban area with a mix of activities? By this token, most urban areas of the world should be called authentic. Does the creation of authenticity in urbanism mean the success of a development scheme to sell all of its units to customers? By this token, many commercial development schemes that are even labelled as bland and generic would be considered authentic.

Can a quiet street or empty building not be an authentic one? (Figure 4.5). If an architectural masterpiece is derelict or a historic district remains underused, would they be considered inauthentic? The declining centres of many historic cities are the examples of this gap between authenticity and use, whereby the once busy streets have been emptied of their livelihood, but they may still be considered authentic urban environments by all heritage standards. The same can be found in the declining industrial areas on urban peripheries, where factories teemed with large numbers of workers are now rusting away, but still considered to be authentic examples of a particular age and culture. While use is a major contributor to urbanity, it cannot be the sole or even primary indicator of urban authenticity.

## Authenticity and materiality

Locating authenticity in the genius of the artist was rejected by the critics who argued for shifting the emphasis from artist to artwork. This would generate a political character for aesthetics, in opposition to those who had turned politics into an aesthetic performance. Authenticity was found in the object's uniqueness and history, its aesthetic value, and how it represented its society. The object was presumed to have an aura that drew on its singularity and authenticity. But we have seen that every object has its own trajectory, located within a social and historical context. This argues against limiting authenticity to unique objects, fixed characteristics, single origins, and aesthetic appeals. While authenticity is a narrative relationship between an object and an origin, determining that origin in an ever-changing world is just a matter of human construction and convention, finding significance in experience and interpretation.

It may be argued that everything has an origin, and by this token, everything is authentic. At the same time, things do not come into existence from nothing (ex nihilo), and so it would be impossible to find a definite origin for anything, and by this token, nothing is authentic. What this paradox shows is that determining the point of origin for things is a human construction, as things are always changing, evolving, decaying, and emerging in new compositions. Pinpointing one moment in this never-ending process as the origin could be merely arbitrary, or just a way of making sense of the world through simplification. Similarly, everything is made of a large number of other things, and deciding what constituent element can be considered as its origin could be arbitrary or conventional. All objects, therefore, could be both authentic and inauthentic, depending on what we consider to be their origin. It could be an element in their constitution or a moment in their evolution; once selected, it acts as the source upon which an object is judged to be authentic or not. The relationship between the object and that origin becomes a relationship between an appearance and a substance, an outside and an inside. In urbanism, this has been reflected in the debates about the relationship between façade and structure, between construction material and ornamentation, between form and content, and between design and use.

If the origin of objects is selected to be their material, such as wood or concrete, then their authenticity may be judged on that basis, and subsequently, many objects, such as wooden furniture or concrete buildings, can be called authentic with regard to that material. If their form is taken to be their source of identity, such as the design that defines them and gives them shape, then authenticity or inauthenticity of an object would be determined through its form. If content is selected as the core of an object's identity, its functions and its social and historical links become scrutinized in assessing its authenticity. If use becomes the criterion of the space's identity, then the intensity of use is taken to show its urban authenticity. However, form, matter, content, and use are not in competition with each other; they are intertwined, and prioritizing any of them is based on the observers' choice of origin.

Authenticity is a judgement of value based on one or more aspects of something. Objects have many aspects, and our judgement includes deciding which aspect we choose as the focus of assessing their authenticity. As authenticity is an evaluation of the relation between something and an origin, both the aspect of that object and the origin to which we relate it, need to be selected. Furthermore, the judges and the process of judgement are also diverse, as authenticity is a context-specific and observer-dependent value judgement. These considerations are all important in showing that authenticity is not a fixed and permanent feature of something, but it is the outcome of many processes of judgement and evaluation, with all that this entails. Owing to its highly particular nature, this judgement can vary according to the personal characteristics and social position of the judge, which can change in response to the changing conditions of what is under scrutiny.

# 5 Event

The last chapter explored the implications of shifting from author to object as the source of potential authenticity. This shift opened the possibility of social analysis and critique, rather than limiting the analysis to the personal talents of single individuals who make the objects. The last chapter examined different aspects of physical objects, exploring the implications of their presence, quantity, matter, form, and content for making objects authentic. This chapter shifts the attention from object to event, with the aim of expanding the horizon of investigation. In a broader sense and at different scales, both physical objects and social actions may be considered events in space and time. This chapter explores the city as a pattern of events, which are not just fixed physical objects like buildings and roads but also the range of actions and interactions within and across these physical spaces. This shift from object to event opens the possibility of analysing the city as a dynamic process of unfolding events, taking account of complexity and change. This chapter analyses the problem of event, regularity, and representation, their implications for authenticity, and the way they shape and transform the urban space and society. This chapter also engages with the way events are experienced and reported, thereby problematizing the challenges facing truthfulness and authenticity. In other words, the chapter engages with both the challenges of authenticity and authentication.

## Random events and regular orders

An event is a unique occurrence, a singularity that may or may not fit into the existing patterns. If an event occurs according to identifiable patterns, it may become predictable and somewhat regular. However, as the last three chapters have shown, authenticity has not been sought in regularities and repetitions, but in uniqueness of individuals, objects, and events. The regular life of the urban society, with its restrictions and compromises, has indeed been taken to be the primary expression of inauthenticity (Trilling, 1972). The existentialist creation of the authentic self advocates making something unique and individual, against the homogenizing pressures of the social mould, to make the course of life like a work of art that would not fit into any pre-established

DOI: 10.4324/9781003451778-5

framework. The concept of aura in the artwork suggests that the work is a unique object unlike any other; it is a singular event rather than a repetition like the reproduced objects of mass production. Therefore, in these interpretations, authenticity is closely aligned with uniqueness; authentic objects and persons are considered to be singular events, rather than regular occurrences.

The singularity of the authentic event is placed in contrast to the generality of patterns and laws. By standing outside these regularities, an authentic event is simultaneously seen as an expression of freedom and emancipation and a critique of the established order, displaying a tension between order and event. If the event is seen as a random occurrence, it is taken to be at odds with order and regularity, and if it cannot fit into any familiar patterns, it may be seen as unruly, and potentially disruptive and dangerous. Science, religion, philosophy, and art have long been attempts at finding and creating meaningful patterns that would make sense of what appeared to humans as random events.

An early attempt at making sense of events may be found in the birthplace of cities, ancient Mesopotamia, where from the beginning of the second millennium BCE a new profession emerged that specialized in listing all the unusual events, taken to be messages from gods as warnings about the future (Bottéro, 2000). These practitioners meticulously and extensively recorded the movements of stars and planets, the appearances of humans and animals at birth and through life, and all kinds of other occurrences. These long lists (amounting to 2,000 lines in one case) were tabulated with future events. If it rained on the day of the god's festival, it meant that the town's god was angry, or if a man had a flushed face, it indicated the impending death of his elder brother (Bottéro, 2000:45–48). Through long-term observation, recording, and cross-tabulation of events, they were looking for causal links between unusual events, and discovering patterns that could generate meaning and certainty. This system of divination and astrology has been at the root of modern-day horoscopes. It has also been at the root of the scientific methods that emerged in Mesopotamia and developed since then, as it initiated systematic observation, recording, and making connections between what seemed to be random events (Bottéro, 2000; Madanipour, 2007).

The evidence for making connections and creating orders out of events goes even further back in time to an area that has cautiously been called the birthplace of art. One of the oldest works of art in the world, from more than 70,000 years ago, was discovered in South Africa and is now in a Johannesburg museum, comprised of abstract geometric patterns on the sides of two small ochre stones (Figure 5.1). It has been interpreted as the earliest evidence of symbolic thought, pushing back the history of art long before the European cave paintings. The regularity of the geometric form shows a desire for the generation of some form of abstract order that would represent the world. Geometric patterns abound in the natural world, and it may well be that these stones show an attempt at imitation, but the fact that early humans selected geometrical forms for their image-making shows an aesthetic

*Figure 5.1* One of the oldest works of art in the world, and the earliest example of symbolic thought in homo sapiens, shows an abstract geometric order (Johannesburg, South Africa)

choice, indicating a desire for regularity and order. It shows seeing patterns and creating patterns as an early human characteristic.

The relationship between events and order finds a powerful and long-lasting philosophical expression in Plato (1997). As we saw in the discussion of form in Chapter 4, Plato divided the world into two levels: the sensible, which was the world of events, and the intelligible, which was the higher world of eternal ideas. When faced with a world of events, with all its apparent randomness and diversity, Socrates and Plato look for permanence and certainty, which they find in an ordered higher realm of primordial ideas. In response, sceptics and empiricists like David Hume have questioned the existence of these higher patterns, arguing that we make sense of events by making associations between them (Morris and Brown, 2022). According to Ilya Prigogine, a Nobel laureate chemist, this law-event dualism is a perennial feature of the history of western thought. For Prigogine, laws described 'continuous unfolding', 'intelligibility', 'deterministic predictions', and permanence, while events implied 'arbitrariness', 'discontinuities, probabilities and irreversible evolution'; so he argued that we should accept that 'we live in a dual universe, whose description involves both laws and events, certitudes and probabilities' (Quoted in Coveney and Highfield, 1991:16).

A generation after Prigogine, Kip Thorne (2013), a Nobel prize-winning physicist, charts this distinction but offers a different description of the relationship between laws and events: rather than seeing them as oppositional, they are described as different images of the same thing. As Thorne argues, the relationship between event and order can be described in two ways. For those who live through events, it is a trajectory of life as it unfolds. It is their lived experience with all its changes and meanings. From a first-person view, it may unfold like a novel, but, from the third-person view, all the events of this life, and all other lives, are already there. For scientists like Thorne, who imagine standing outside and mapping the events that make up the world, the entire universe and everything within it is made of events in a four-dimensional fabric of spacetime (Thorne, 2013). By moving between a first-person perspective and third-person perspective, what seems to be random is given an overall shape, an order that can make sense. By envisaging this overall order to the universe, randomness is almost abolished at one level, while it remains experienced at another.

This may be the case even when an event is seen as a critique of the status quo, which can be found in the debates on innovation and techno-economic change. The literature on innovation sees disruptive events as a potential for positive change, whereby the normal course of events with its tired and repetitive behaviour is challenged, enabling the change of direction. But these disruptions are part of the system, rather than a shock from the outside. A prominent perspective in this literature, developed by Schumpeter (2003) in the middle of the last century, draws on biology, where genetic mutation can lead to better adjustment to the environment and an improved chance of survival. Rather than a risk, the singular disruptive event is seen as a potential strengthening of the organism for survival. Innovation

> incessantly revolutionizes the economic structure *from within*, incessantly destroying the old one, incessantly creating a new one. This process of Creative Destruction is the essential fact about capitalism. It is what capitalism consists in and what every capitalist concern has got to live in
>
> (Schumpeter, 2003:83)

Creative destruction removes the obsolete structures, paving the way for innovation, which comes from 'the new consumers' goods, the new methods of production or transportation, the new markets, the new forms of industrial organization' (Schumpeter, 2003:83). However, this celebration of disruptive events takes place within the regularity of an economic system that is not bound to change through these disruptions. The event is part of a pattern of regularity, rather than a fundamental point of change.

In contrast, the singularity of the event can become a challenge against the entire order. Michel Foucault (1981) uses the dualism of order and event in social criticism, arguing that the ordering processes, especially in the form of discourse, are inherently the processes of exclusion. He refers to events as the chance element, which the order of discourse tries to bring under control. His hypothesis, presented in his inaugural lecture at the Collège de France, was that

> in every society the production of discourse is at once controlled, selected, organised and redistributed by a certain number of procedures whose role is to ward off its powers and dangers, to gain mastery over its chance events, to evade its ponderous, formidable materiality.
>
> (Foucault, 1981:52)

He identifies several procedures that are external and internal to discourse, which are used to tame the unpredictability of chance and limit the potential of events for undermining the status quo. Foucault does not use the language of authenticity, and he is indeed critical of placing an emphasis on individual agency. But his critique of the limits imposed by structures, institutions, and discourses on human action and thought, along with his emphasis on the event as a chance element subject to these limitation, shows that he too values uniqueness and singularity, with their potential for freedom from imposed social orders. This becomes more evident in his later work on the care of the self (Foucault, 1986). In an interview later in life, he declares that 'I believe in the freedom of people. To the same situation, people react in very different ways' (Quoted in Martin, 1988:14).

However, as regularity is the realm of general concepts when singular events are interpreted as being authentic, they are brought under an umbrella term, hence assigning a general pattern to them. This shows a paradox in which a class of unique events is envisaged under the universal title of authenticity. Moreover, our earlier questions about the role of quantity in authenticity (Chapter 4) may equally be applied to events. In reading the world as a series of events, we are faced with the possibility that every event is a unique and singular one. Two events may be similar, but almost never fully identical; they are grouped together for their similarities while downplaying their differences. If only judged by the randomness and uniqueness of events, the possibility of authenticity diminishes radically, as every event, like every object and every person, is a single nonrepeatable occurrence. But it is never a single occurrence in a void, as it is always part of a series of unfolding events, before and after it. An event is never a complete rupture with no history, as it is always linked in some way to what went on before, and carries the seeds of what might happen afterwards. In everyday discourse, however, an event can be commonly understood as a self-contained occurrence, which can then be used as a point of reference, an origin, from which some authentic object or event could arise.

### From major events to everyday events

For Alain Badiou (2005), an event is an unusual rupture that changes the normal course of life, which has major consequences, paving the way for truth. 'A truth is solely constituted by rupturing with the order which supports it, never as an effect of that order. I have named this type of rupture which opens up truths 'the event'' (Badiou, 2005:xii). Badiou's events are turning points that are not shaped by their conditions, but they go beyond these conditions to change things in radical ways. As such, they are universal, but they are also singular as found in the four domains of life, that is, love, art, science, and politics. These events can become manifest in: falling in love, emergence of a new movement in the arts, a major scientific discovery, and a transformative political revolution (Badiou, 2022). After these events, life is no longer the same as before. For Badiou (2005:327), truth is fidelity to these events: 'truth is always that which makes a hole in a knowledge'. As he explains, 'a truth groups together all the terms of the situation which are positively connected to the event' (p.335). The events are radical ruptures, and accordingly, this truth is infinite and distinguishable from knowledge, which is finite (p.333).

Fidelity to the event provides the possibility of truth, that is, personal commitment to working out the consequences of a major event (Badiou, 2005). For Badiou, therefore, a major event functions as an origin, around which a believable account can be developed. Badiou accuses others for their nostalgia for a deleted origin, but in privileging the major events, he seems to have found those origins, to which he argues to remain faithful. These major events act as the origins of a form of authenticity, which is not about knowledge, but about faith. It shows some similarity to defining knowledge as justified belief (Madanipour, 2011), which aims to avoid the pitfalls of epistemology and accept the possibility of failure in knowledge, hence limiting it to belief.

Badiou (2009:381) criticizes Sartre for ignoring the chance-events: 'Sartre was still concerned with the auto-unification of the transcendental. He did not expose the subject to the chance of a pure Outside. One of the names of this Outside is 'event''. Badiou (2009) distinguishes his concept of event from Deleuze's concept of event, which he criticizes for its vitalism and its unitary character. He accuses Deleuze that 'he opts for fate. The event is not the chance-laden passage from one state of affairs to another. It is the immanent mark of the One-result of all becomings' (Badiou, 2009:384). He also criticizes the phenomenologists and existentialists for their religiosity. In Husserl and Heidegger, authenticity is not rooted in a real event, but in nostalgia for a lost event:

> Phenomenology, in its German variant, is indisputably haunted by religion. This probably stems from the motif of a lost authenticity, of a forgetting of the true Life, of a deleted origin—a theme that traverses all of Heidegger's writings, but which is also fully established in Husserl's *Krisis*.
>
> (Badiou, 2009:516)

However, Badiou's own emphasis on the event as an uncaused cause, that is, something that is not part of a chain of causation, but it is a major transformative occurrence that starts a new causal chain, and his emphasis on the universality and infinity of the truth are not free of religious connotations.

Slavoj Zizek follows Alain Badiou in adopting the concept of the event as a major rupture that goes beyond its conditions of possibility. According to Zizek (2014:179), 'In capitalism, where things have to change to remain the same, the true Event would have been to transform the very principle of change'. The change may be in the change of perspective or a change in material reality, that is, in the way we see things or the way things are changed in themselves. Therefore, an event

> at its purest and most minimal: something shocking, out of joint, that appears to happen all of a sudden and interrupts the usual flow of things; something that emerges seemingly out of nowhere, without discernible causes, an appearance without solid being as its foundation.
>
> (Zizek, 2014:2)

This analysis does not look for a chain of causations, as the various forms of determinism, from Marxism to scientism, had held, while also aiming to avoid the liberal theory of autonomy, as well as the existentialist theory of authenticity. Events are merely chance occurrences that are not caused by or based on anything discernible, but their occurrence causes transformative change.

However, focusing on major, life-changing events runs the risk of ignoring the many thousands of smaller events that make the course of life. While big ruptures do happen and can change the world in significant ways, most of the social processes go through slower, less visible changes, which together may amount to substantial transformation. These may happen in physical processes of evolution and change, in individual psyches and biographies, as well in social processes and economic and political happenings. The ancient heap paradox questions: if we take one grain at a time from a heap, when does it stop being a heap? In the analysis of events, we can ask the reverse question: at what point does the continuous addition of one grain to another grain start forming a heap? In other words, the accumulation of small events can have major consequences without being visible as a sudden rupture (Figure 5.2).

More importantly, each small event can have an impact on a single life trajectory, so it does not make sense to dismiss it as inconsequential and unimportant. Privileging significant events for being transformative should not take the attention away from the impact of banal, everyday events on individuals and groups. The implications of this change of perspective, from major events to all events, for the analysis of authenticity would be noticeable. A theoretical framework which sees major ruptures as the reference points for truth is not far from the notion of the event as the origin of authenticity. Badiou's theory of event and truth does not mention the word authenticity, but it can be seen as a version of the idea of authenticity. It seeks to

*Figure 5.2* Focusing on major events could miss the significance of the everyday events that could shape and transform lives (Istanbul, Turkey)

establish a truthful relation between a course of life and a major event, which can be in art, love, science, or politics. However, the argument that needs to be made against this theory is that these reference points do not emerge in a void and they are not limited to major events. Large and small events, sudden or cumulative, may all be adopted as points of reference on which people may construct their own narratives and trajectories, based on their own particular circumstances.

The small and unimportant objects and events can play a major role in the development of thoughts and behaviours, which may remain hidden and go unnoticed in the analytical processes. In urbanism, this can be seen in the longstanding practice of mental mapping. When Kevin Lynch (1960) used mental mapping to develop his five memorable principles of urban space and design, his focus was on the key elements that could be derived from individual mental maps. His method was looking for patterns in a range of events, which were individual memories of walking round the block and recorded as personal maps. By overlaying many mental maps and prioritizing the major points of reference as he saw them, he developed an influential theory of the city. For him, the monuments, paths, nodes, edges, and districts provided the patterns through which to make sense of the events that were recorded as the experience of the urban space.

However, in this theorization, what was edited out were the small details, unimportant objects, and singular experiences, which can indeed be much more important in memories of the city and navigating the urban space. A tree, a shop sign, a difference in the colour of the pavement, a place of gathering at a street corner, can all be the points of reference for reading the city. Lynch develops a theory of space by privileging the significant and the large-scale objects while ignoring the singular, the unimportant, and the small scale, which may have a much more important impact on a person's memories and future actions. In mapping the memories of a journey around the block, these are small events that may go unnoticed, but each may have significant impacts on navigating the city and remembering its parts.

## City of habit

Beyond the images that are on display and the spaces that are inhabited, the events of the city may be analysed as a set of behaviours, which are a combination of habitual and spontaneous events, of regularities and random events. A large city, especially in its points of concentration such as railway stations, airports, and central districts, may be overcrowded, with people who move in all different directions. This may puzzle the unfamiliar visitor, who feels the energy of the crowd, but might find it confusing and think of the movement of people as random, without an identifiable pattern. However, there are many individuals in that crowd who would be following their daily habits and routines, repeating what they normally do. To them, everything in their behaviour is routine rather than random. Cities tend to be spatially structured and socially organized by these daily routines. From the outside, the city may appear as a random coexistence of unrelated events, but from inside, the city is structured by a social and physical infrastructure, and by individual habitual trajectories (Figure 5.3).

A life lived through habits and routines within a tightly organized, socio-spatial structure has often been considered to be the epitome of inauthenticity. As cities re-emerged in Europe in the late medieval period, the necessity of high-density living within the city walls required new forms of social coordination. After the emergence of the modern capitalist industrial city in Europe, these forms of social coordination were transformed into a pattern of laws, norms, habits, and routines that have given its inhabitants the frameworks for living together in dense urban environments. At the same time, this rational order has been criticized for being a straitjacket that leads to social alienation and inauthenticity. A desire to rediscover an imagined lost originality lies at the heart of many narratives of authenticity. By this token, the urban dwellers' habits may be seen as the hallmarks of inauthenticity, a sign of their subordination to an alienating social order.

*Figure 5.3* The apparent disorder of a crowd may hide the many habitual routines that structure the city (Tokyo, Japan)

*Transformation of habits*

In a fascinating study, Keith Thomas (2020) charts the transformation of manners in early modern England, during the period of sixteenth to eighteenth centuries. This was the beginning of the European voyages of discovery, which ended in the colonization of much of the world. In their encounters with this world, a question that the writers in this period were concerned about was how to be civilized, who the civilized nations were, and how they differed from the barbarian others. Through the construction of an image of others, the explorers and colonialists helped develop a sense of identity about how society should be organized and how social life should be lived (Thomas, 2020:9). It was through these encounters with others throughout history that a sense of the self was created for European societies (Frankopan, 2015).

Otherness was created through space, in other regions and countries, and through time, in an imagined encounter with the people of the past. For the modern observers, the Middle Ages was a period of incivility, thankfully swept away by the Renaissance. Writing in the 1930s, Norbert Elias (1978) had thought that the medieval period was 'a time of unchecked impulses', lack of self-control, with 'childlike oscillations of mood', careless 'about the bodily functions and a disposition to spontaneous violence' (Thomas, 2020:19). With the dawn of the Renaissance, Elias had suggested, people went through a 'civilizing process', which involved the self-regulation of 'bodily drives and emotional impulses' (Thomas, 2020:19). A third source of otherness was in social hierarchies within the country, where the manners of the court and aristocracy were thought to be superior in civility and urbanity to those in the provinces and rural areas.

However, Thomas argues that these differences were changes in conventions, rather than becoming more civilized through increased self-control. With the emergence of 'a public realm of sociability', the subject of manners and the skills that were needed for peaceful and effective social interaction became a topic of concern (Thomas, 2020:21–22). Civility, as both David Hume and Lord Chesterfield had argued, would involve 'sacrificing one's own self-love to that of others' (p.26). Politeness required being pleasant and elegant, avoiding all forms of 'dissent and angularity' (p.29). Its manifestations in bodily gestures, posture, speech, and manners were all marks of social distinction, based on defining civility as emotional self-control. Urbanity, which originally referred to 'pleasantness of speech' became more generally known as 'a synonym for civility and courteous behaviour' (p.58). The growth of politeness, civility, and urbanity reflected the development of a new social order with its social hierarchies, and the more they grew the more they became the subject of transgression.

This analysis of the pursuit of civility shows how social habits in England and other European countries changed over a period. What are the implications for our analysis of authenticity? The early romantic writers thought that the drive for politeness had led to the loss of authenticity: as a Scottish poet wrote in 1776, 'The conversation of the common people, though not so smooth, nor so pleasing as that of the better sort, has more of the wilderness and strong expression of nature. The common people speak and look what they think' (James Beattie, quoted in Thomas, 2020:106). Should we conclude that as the codes of social conduct grew and became more complex, society became more inauthentic? Were people in urban areas and higher echelons of society, who observed these codes, more inauthentic, while the rural populations and the lower income groups lived an authentic life? If people were increasingly expected to behave in certain ways, were they no longer expressing themselves and therefore becoming ever more inauthentic? Was it not possible to be polite and authentic at the same time? Was it not possible to be authentic and still observe certain forms of social decency, such as the performance of bodily functions or behaviour towards strangers?

As Thomas (2020:20) argues, 'There never has been a society without any conventions about proper modes of bodily comportment and social interaction', so 'what changed in the early modern period was not the degree of personal self-control, but the prevailing conventions about how and when emotions should be publicly expressed'. If the expression of emotions was the axis around which the new codes of manner revolved, civility and authenticity would become the two opposite ends of a spectrum. Civility, politeness, and urbanity involved developing the habits of emotional self-control, while authenticity would involve dismissing these norms, thereby releasing these emotions, and expressing them more freely in action and behaviour. In some ways, the rise of eccentricity in the urban middle classes can be seen as an expression of this defiance against the social norms, while inhabiting in a comfortable milieu that those social norms and power structures have provided.

This suggests that authenticity, in whatever way defined, is a critique of social habits and conventions, describing a state of being that is an alternative to the status quo. And yet striving to be authentic can be another social convention, not unlike the conventions that it seems to be out to break. When the notion of authenticity emerged with the romantic movement, it was part of the development of a new set of cultural patterns, which were defined through the negation of the status quo, which had forced social life into tight rational frameworks. In developing authenticity as a critique of these rational orders, a new culture of nonconformity was developed, which in a sense was another pathway to conformity. Breaking the conventions becomes a convention itself, a social performance through which individuals define their identity through differences rather than similarities, but in doing so they establish new structures of similarity (Madanipour, 2013). Examples of architecture and urbanism would be the desire to design and build buildings that are different from others. It is a claim to a separate identity that is established through difference. However, when a number of such buildings accumulate in a city, and around the world, being different becomes a habit, a common norm from which it is difficult to escape. Similarity, rather than difference, becomes the outcome of the bid to be different.

### *The second nature*

It has been suggested that Norbert Elias, with his pioneering work on social habits and manners, and with his relational thinking, influenced Pierre Bourdieu (Dépelteau, 2013). Striking similarities have been found between their three key concepts: habitus, field/figuration, and power/capital (Paulle et al, 2012). In particular, the concept of habitus is relevant for our investigation, with which they meant 'a system of acquired dispositions', similar to Aristotle's concept of 'second nature', so as 'to emphasize the importance of taken-for-granted ways of perceiving, thinking, and acting on the part of (more or less) competent actors immersed in their everyday practices' (Paulle et al, 2012:71).

The ideas of habitus and second nature indicate a degree of longevity in social habits, developed as accumulated dispositions, which undermines the idea of the self-knowing subject who is detached from the world. According to Bourdieu (2000:138), 'social agents are endowed with habitus, inscribed in their bodies by past experiences'. This allows Bourdieu to avoid the dichotomy of individual and society by locating the individual in a social context: 'Habitus understood as an individual or a socialized biological body, or as the social, biologically individuated through incarnation in a body, is collective, or transindividual' (Bourdieu, 2000:157). When habitus was displaced, it could lead to conflict and contradiction. As Elias argued, 'the pretension to be what one is not [leads to] insecurity of taste and conduct, "vulgarity"' (quoted in Paulle et al, 2012:71). Moreover, due to its inscription on the body, habitus cannot be easily removed through the talking cure of psychoanalysis. Bourdieu (2000:172) argues that 'While making things explicit can help, only a thoroughgoing process of countertraining, involving repeated exercises, can, like an athlete's training, durably transform habitus'.

In this analysis, however, there can be little room for authenticity in individual behaviour and character, as it denies individual uniqueness, and describes the human being as a reflection of society. Bourdieu (2000:132) condemns 'the "personalist" belief in the uniqueness of the person', as 'the basis of the scientifically disastrous opposition between the individual and society'. For him, 'Personalism' is 'a condensed form of all the theoretical postures – mentalism, spiritualism, individualism, etc.' and it is 'the main obstacle to the construction of a scientific vision of the human being' (p.132). The quest for social distinction and truthfulness, with which the concept of authenticity appears to be aligned, leads to the complex web of structures, functions, and conventions of the clusters of habitus that make up different social fields. For example, it is the artistic field that shapes the aesthetic dispositions and defines the manners, styles, and schools of art (p.100).

The notion of second nature goes back to Hegel, who links it with the development of an ethical character. If the first nature is the way people are born, the second nature is what they become by living and learning in society, when the ethical becomes 'simply identical with the reality of individuals' (Hegel, 2001, §151). The first nature is fused with the second nature, the two natures turning into one: 'This is the custom, which as a second nature has been substituted for the original and merely natural will, and has become the very soul, meaning, and reality of one's daily life' (Hegel, 2001, §151). The second nature is an outcome of education, which has the potential to transform humans from natural to social beings:

> Pedagogy is the art of making men ethical. It looks upon man as natural, and points out the way in which he is to be born again. His first nature must be converted into a second spiritual nature, in such a manner that the spiritual becomes in him a habit.
>
> (Hegel, 2001, §151)

This conversion is only possible if people learn to follow an appropriate ethical path, as a result of which good habits become rooted in people, turning them into ethical beings, as distinctive from natural beings.

Hegel's notion of second nature clearly follows Aristotle (1991), who argues that 'habits, when exercised, increase and grow' (*Problems*, vi, §26) and through frequent repetition, they become a person's disposition and character (*Eudemian Ethics*, ii, §2), 'a second nature' (*Problems*, vi, §26). The best habits are about the practice of moderation (*Eudemian Ethics*, ii, §5), which develop through education and training from childhood. Through these habits of moderation, a person's character develops and leads to moral excellence: 'moral excellence comes about as a result of habit', which we learn by doing, as 'none of the moral excellences arises in us by nature' (*Nicomachean Ethics*, ii, §1). The highest aim is to achieve happiness, which is 'living in accordance with the excellences' (*Magna Moralia*, i, §4). By learning and practising moderation, good habits are formed, growing to become second nature, and leading to moral excellence, happiness, and flourishing. This ancient Aristotelian model is based on moderation, self-control, social engagement, and character development, which seems to be the complete opposite of those interpretations of the modern concept of authenticity that are opposed to society, advocate a return to nature, and assert individual exceptionalism. While ancient ethical wisdom emphasized moderation and avoidance of excess, the modern romantic attitude is built on pushing the boundaries and seeking excess and infinity (Badiou, 2005; 2009).

The Hegelian concept of second nature, which lies at the heart of Bourdieu's model of habitus, draws on Aristotle's notion of second nature, albeit with noticeable differences. In Aristotle, it meant the process of human flourishing in which the person's life matches their predetermined natural purpose, while in Hegel it indicated the process of self-creation through social engagement, linking the material and ethical life (Demir, 2020). In these concepts, the origin on which any sense of authenticity may be based is not in the biological constitution of the human being, but in a second nature that has evolved through the course of life in society. Social habits, far from being the enemy of authenticity, lie at the heart of a social interpretation of authenticity. Habits are forms of regularity that shape much of social life, so the question becomes whether authenticity may be found in this regularity or through the ruptures and forms of resistance.

If personal character and social life are formed through routines and habits, and if these habits become an integral part of life, can habitual life be authentic at all? And if authenticity is an exercise in going beyond these habits and routines, is it possible to find a place that is completely devoid of any habit? As we have seen, all social behaviour is mediated through common frameworks such as language and culture and oriented towards intersubjective engagement and social recognition. Authenticity as uniqueness and self-reliance cannot mean abandonment of all habits and complete detachment from these social frameworks, but only an expression of a need for finding a distinctive place among others. By necessity, this distinctive place combines the relations of similarity and difference, rather than being purely an assertion of difference.

## City of performance

The next question is, if social life includes a large dose of performance, can authenticity be found in such performances? As already discussed in the last section, the first answer to this question would be about habits and routines which tend to inscribe performances on the body, turning them into integral parts of a character, so that any reference to uniqueness would take us to these internalized habits. An implication of this process on planning and design of urban space would be resorting to tradition, as a social habit, which would reproduce the environment in a stable, repetitive way, albeit with small and gradual innovations. However, as we have seen so far in the book, the quest for authenticity has often been a reaction against this habitual use of traditions, and in favour of using new and original ideas and practices. Meanwhile, as we shall see in Chapter 6, in the discussion of time as the origin of authenticity, returning to the past has largely been seen as an important form of seeking authenticity. Therefore, authenticity is paradoxically sought in both the traditional and the new, in continuity and familiarity as well as in novelty and change, showing the context-dependent notion of authenticity.

A personalized habit may be unique, formed in specific conditions, and thereby a candidate to be considered authentic. As discussed in Chapter 3, people develop and wear different masks in response to different social situations, so that they can control the situation and be seen by others in a good light (Goffman, 1969). As these performative efforts become consolidated into people's social persona and personal habits, the performance may even be considered more authentic than what lies beneath (Bourdieu, 2000; Butler, 1990; Zizek, 2006). The question now is about the possibility of authenticity in performances that are purposeful, rather than automatic and integral to character. These purposeful performances show an active agent at work, rather than habit and routine. Can there be a stronger element of authenticity here? We may mention two such performances: by the speakers of a language and by performing artists.

Performativity in language refers to the active role that words can play in generating events. In speech act theory, performative sentences are those that do not just describe the phenomena but also create them (Austin, 1962; Green, 2021). Performatives are 'declarations' that create social and institutional facts, utterances in which 'saying something makes it true' (Searle, 1995:34; 1999:115). An example is when the person in charge of conducting a marriage announces that 'I now declare you husband and wife', thereby producing a new social reality by this utterance. Together with non-linguistic devices, such as bodily gestures (Goffman, 1969), performativity is used to show how ideas, practices, and identities are socially constructed. Therefore, performance is an act that constitutes social reality. As Zizek (2006:32) argues, following Lacan, such performances may be more authentic than what is assumed to be inside the mind. In other words, the authenticity of social reality is established through performative action, rather than mental processes and subjective reflexivity. In urbanism, this seems to suggest that the public environment, what lies outside and is visible and accessible to others, is the authentic city, rather than what lies behind closed doors. It would also mean that there is no predetermined essence to the city, and it is what is made through the actions of the citizens and others.

A problem with this interpretation is that performativity cannot be detached from its context, which is social and historical. A performance is not suddenly appearing out of nowhere, but it is part of a chain of events. It has a trajectory which may not have determined it but without which it cannot be understood. The social reality that is created by an utterance, a performance, or a construction can only make sense when located in its web of concepts and practices, their evolution and transformation, and the common frameworks that link the speaker to the listener, the performer to the audience. The physical environment may be new, but it is not created in a void by people who had no idea of what a city is, but it draws on a series of ideas, practices, skills, and memories. The physical environment is one of the non-linguistic ingredients that is often neglected in the analysis of performativity. It contributes to the constitution of these socially constructed meanings, as the backdrop to the performance and as an integral part of the performance. In other words, as we have argued, the inside and the outside, the embodied and the embedded are in continuous interaction, and any search for authenticity would have to engage with this interaction.

The discussion of authenticity in theatrical performance finds expression in several relationships: the relationship between performer and character, between performance and text, between performers and audience, and the relationship between performance and space. In the relationship between the performer and the character, the problem of authenticity lies in the paradox of acting, which by definition is a copy of an imaginary character, a reenactment of someone else, rather than performing the actor's self. Actors are required to pretend to be someone other than their own selves, so how can there be a possibility of authenticity here at all? How can this be unique, as many actors may have played the same role. Should the actors be truthful to the character, thereby immersing themselves fully and faithfully in the character, or should they be their own selves that recount a story, keeping a distance between the self and the character that is visible to the spectators? In relation to the text, which may have been produced in a different age and culture, how far should the performance try to recreate the atmosphere of the past, or should it retell the story in a contemporary idiom and environment? How should the actors relate to the audience, delivering a performance to the audience or engaging with them in an interactive, shared experience? Should the stage be designed such that it appears to be a realistic reproduction of the play's original time and space, or should it be freed from such imitation? How can an authentic theatre be created? These are the questions that have faced researchers and practitioners for a long time (Brook, 1990; Gamel, 2013; Hannah, 2019; Schulze, 2017; Walmsley, 2013).

Authenticity remains an important concern for theatrical performance (Walker, 2019). Researchers have produced different taxonomies for the types of authenticity in theatre. In one attempt in relation to playing the works of ancient drama, several types of authenticity are identified: nominal/historical, expressive, processual, structural, inductive, and critical (Gamel, 2013). In nominal or historical authenticity, the concern is to recreate the play as it was intended and received at the time of its creation. Expressive authenticity, in contrast, reflects the performer's own beliefs and values. Processual authenticity is more democratic, involving all the artists. Structural authenticity is found in community theatre, involving all the stakeholders including the artists and the audience as part of the local community. Inductive authenticity is when the performance is not faithful to the ancient script but can be innovative and subversive. Critical authenticity engages with the social goals of the performers and how the work has been produced (Gamel, 2013). Another attempt focuses on the experience that is created through performance, listing intimate, immersive, and documentary forms of authenticity in theatre (Schulze, 2017).

The substantive core of theatre has ultimately been identified as communication between the actor and the audience (Walker, 2019). This is expressed by the famous theatre director Peter Brook (1990), who has his own classification of theatre: deadly, holy, rough, and immediate, describing commercial, elite, popular, and his own form of theatre. For Brooks, what matters is the communication, which can be cut down to its bare form, cleansed of all the ornamental paraphernalia such as red curtains, spotlights, darkness, etc. He argues that, 'a beautiful place may never bring about explosion of life, while a haphazard hall may be a tremendous meeting place; this is the mystery of the theatre' (Brook, 1990:73). As he puts it,

> I can take any empty space and call it a bare stage. A man walks across this empty space whilst someone else is watching him, and this is all that is needed for an act of theatre to be engaged.
>
> (Brook, 1990:11)

In theatre architecture, a gap emerged between art and architecture in the twentieth-century avant-garde. While the artists were increasingly focused on 'the fleeting acts', architecture continued with its tradition of the design and construction of buildings, as expressed in 'a new classical modernism' (Hannah, 2019:xv). In some situations, this gap between the act and the object could be bridged by bringing them together, but even then it seemed that this joint presence would not lead to mutual influence.

There is always a substantive difference between a performance and the space in which it unfolds. The performance is timebound, taking many shapes and characters, according to each performance. The space is often stable, open for adjustment to these changes, but within a degree of permanence in infrastructure. The same space may be used by dancers, actors, and musicians on different occasions. This would require a change of interior space, between different shows and even during a performance, while the theatre building and its basic layout remain the same. The performers may need to make changes to the stage to fit their purpose, but they are also constrained by some of the fixed properties of the stage. The relationship between space and event reflects compromise and adjustment, a gap between solidity and fluidity, as well as a division of labour between the designer and the performer. This is why Brook (1990:74) recommends to theatre designers to

*Figure 5.4* The urban life combines performativity and spontaneity, framed spaces and unplanned actions, as a marketplace demonstrates (Isfahan, Iran)

work like scene designers, 'moving scraps of cardboard by intuition, than if he builds his model from a plan, prepared with compass and ruler'.

There are significant parallels between theatre design and city design, especially if public life is considered a performance. The urban space is the stage on which the drama of public life unfolds. However, there are serious limitations in applying the dramaturgical model of masks and performances to social life, as there are many occasions that are spontaneous, accidental, naked, and without a socially recognized mask (Figure 5.4). Urban life and urban space are full of these social and spatial events, which are not designed, or that do not function as intended and planned. There are also serious limitations in applying the concepts of the theatre stage to urban space, as the urban space must accommodate a vast range of unpredictable events, compared to the more limited range of institutionalized and choreographed events in theatrical performance, organized according to a script. Authenticity in theatre may lie in the expression and communication of meaning, while authenticity in the city may lie in the interplay of regularity and randomness, recurring habits and spontaneous happenings, and solid materiality and ephemeral events, both inside and outside.

## City of fleeting images

The widespread use of images in the contemporary city has turned urban life into an intense visual experience, a continuously unfolding collection of visual events. At some point, the proliferation of imagery was thought to be a sign of the information society, creating and communicating information in many new ways needed to live in a complex society (Madanipour, 2011). The proliferation of imagery was also seen as creating a society in which spectacle prevailed at the service of capitalism (Debord, 1994). It signified major changes in technology, economy, and society, whereby industrial cities and routinized ways of life were replaced by services, digital technologies, and diversified life routines (Madanipour, 2011). The question for our inquiry is whether any sense of authenticity may be found in this highly visual culture.

Living in large urban areas involves continuous travel through long distances, turning the urban dwellers into permanent travellers and the experience of the city into fleeting moments. This limits the travellers' physical contact with the city, turning their urban experience into what may seem like watching a film, viewing a long and frequently interrupted sequence of rapidly passing images. As Georg Simmel (1950) argued long ago, the urban dweller becomes overwhelmed with this overload of information, resulting in a loss of sensitivity and an attitude of indifference to the urban environment. To counteract this reduced level of sensitivity, the figure of flaneur has been championed as someone who engages with the urban scene, rather than passing through at high speed or with a blasé attitude (Arendt, 1969; Benjamin, 1969). By walking slowly and aimlessly, the flaneur is expected to experience the city in a more engaged way. And yet, even the flaneur's experience is often skin-deep, looking for aesthetic pleasure in an encounter with surfaces, rather than what lies behind and beneath. The experience of the urban dweller, at high or slow speed, becomes an exercise in surfing the surfaces of the city. The city is experienced as a series of visual events, some disturbing, some interesting, some informative, some remembered later, and some completely ignored or soon forgotten (Madanipour, forthcoming).

The city becomes a screen on which the visual culture is performed, the stage on which an endless array of designed and produced images is on display, increasingly larger, louder, brighter, and more colourful (Figure 5.5). They are not random images, produced without a need or purpose, even if they appear to be so to the urban traveller. However, their purposes may

*Figure 5.5* The visual culture is displayed on ever-larger, ever-present digital screens (Stockholm, Sweden)

be complex, overlapping, and misunderstood. Together they create a visual atmosphere into which the urban dwellers are socialized, getting used to navigating them, and learning the skills of reading the signs. With the decline of manufacturing industries in former industrial cities, the rise of service-based and consumption-based, globalized economies, and with the development of new digital technologies, images have become an integral part of life in the city. As the strictly structured patterns of life in the industrial city gave way to the multiple and flexible routines of 24-hour cities and service-based economies, the visual culture of the city finds new, overlapping functions.

The making and displaying of images can be used for the purposes of information, animation, and embellishment. One of the purposes of these images in public spaces is to provide information, as displayed in public safety signs, tourist maps, house numbers and name plates, street names, signs of institutions and companies, shop signs, digital and paper news bulletins, information about cultural events, weather forecasts, and air temperature. Another purpose is to animate the city, using colour, electricity, and light to change the atmosphere, especially in dark winter nights. Illuminating the city has long been a visual method of bringing delight and festivity to the city, as well as the more functional role of facilitating travel at night and navigating the city, the economic role of extending the working day, the social role of brightening the home life, and the political role of exerting more control over the life of the city. Images are used for embellishment at the personal level, such as in items of clothing, jewellery, makeup, and tattoos, and at the city level through colour, light, and ornamentation.

The making and displaying of images are used to monitor, influence, and control behaviour. A vast and ever-growing network of digital cameras and technologies of data capture and face recognition are installed on the roads, on car dashboards and body cameras, at the entrances and inside the buildings, at the cash machines, on computers and phones, and almost every conceivable surface. They capture and create images that monitor behaviour in public and private spaces, blurring the boundaries between privacy and publicness. Images in advertising aim at influencing behaviour, attempting to persuade us to purchase goods and products, and competing for the attention of the urban traveller more than any other imagery. The advertising images are a constituent part of an economy of attention, in which attention finds monetary value, but by their frequency and intensity, and in the incessant passage of images from in front of the urban dweller's eyes, these images form an ever-present, overwhelming background noise. The images are also installed to control behaviour, such as traffic signs, that determine where the drivers can or cannot go, and a large number of other signs that instruct the urban travellers in their journeys across the city.

The urban imagery sets up a cultural framework that aims to shape urban life. Since the invention of mechanical clocks, public clocks have set and regulated the rhythms of the city, replacing the natural rhythms of day and night. Historical traces and the signs that refer to them, such as memorial plaques, attempt to construct a collective memory. Statues, posters, and monuments symbolize political and economic power, as do the shape, façade, entrance, height, and scale of buildings that boast about wealth and significance, expecting to instigate awe in the urban traveller. The urban events and their imagery, as in public art, festivals, and parades, help construct a cultural order, while other images, such as graffiti and protest banners, question and challenge that cultural order.

It is difficult to see how the concept of authenticity can be applied to any of these images when encountering them. They all look like everyday messages that compete for the attention of the urban traveller, and none appear to have the singularity that would serve as an origin for an experience. Can a traffic sign or an advertising poster be found to be authentic in any sense? If they construct a city of surfaces, how can a surface, an outside, be the basis on which authenticity may emerge? Is it not an inside that makes something authentic? It would partly depend on how we define authenticity.

Items in this flurry of urban imagery could be called authentic if we have selected authority and functionality as the points of reference for their presence, and if they perform the function for which they have been produced. An authentic traffic sign would be one that is installed by the relevant public authority responsible for traffic management, that is similar in style and colour to other such messages, that provides accurate information, and that is perceived and recognized as such by the urban travellers. It is not a fake traffic sign that is made in the wrong colour and shape, placed in the wrong location, and provides misleading information. In other words, the visual message is an object that is authenticated by virtue of its author, form, content, and experience.

The urban image as a surface is an outside based on some sense of inside, for some purpose and in some intended form. As such it has the potential for being recognized as a form of authenticity as the truthful relationship between something and an origin. But these images construct the rhythms of the city, they are regularities rather than singularities. Meanwhile, the intensity of the city's visuality makes it very hard for any visual sign to perform its intended task in full and reach its level of authentication. The message may be easily missed in the visual hum of the city. Moreover, their original function may be misunderstood or displaced by intention or by mistake, and urban travellers may interpret them in a wide range of alternative ways.

## Between the event and the account

The origin of a claim to authenticity may be an event and the question of how to report it. The current concerns about fake news and fake identities can be seen as an aspect of the problem of authenticity. A narrative of an event is considered to be authentic if it is a true account of that event. This has been a longstanding epistemological concern for science, which has been searching for reliable and accurate accounts of the world. The sceptics and the critics, however, have long questioned the possibility and reliability of truthful accounts. With political polarization, social fragmentation, digital technologies, enlarged networks, and the declining power of traditional media, finding reliable accounts of events has become more difficult than ever before. The same factual information may be interpreted in completely different ways by political groups, with significant implications for the way cities and societies are analysed and managed. The problem of authenticity is also paramount in contemporary visual culture, where it is possible to make and edit images to represent or misrepresent real events. In advertising, the notion of authenticity is widely used, but there is always a concern about the truthfulness of the claims made. In design terms, this has been reflected in the relationship between form and function, and whether an object's appearance represents its functions or these functions are masked by ornamentation.

For long, at least since the invention of photography, an image has been taken to be a truthful reflection of reality. The photo does not lie, it was thought. The camera truthfully represents the real events as they happen. Illustrated magazines and photojournalism could provide reliable information about faraway lands, different cultures, and events, such as wars and natural disasters. The truth of the lost worlds could be recovered through the image (De Cauwer, 2021). The photograph was a truthful record of space and time. A deep sense of trust was established in the honesty of the camera and the authenticity of the image.

There have been many techniques of transforming the image to look better or different. The famous examples of removing the images of high-level Soviet officials after Stalin had eliminated them from the political scene have been used as early forms of fake imagery. Images could also be presented as a truthful account of an event, but, in fact, being staged. For many years, Robert Capa's photograph of the falling soldier was thought to be a moving account of the Spanish Civil War, but its authenticity was doubted in the following decades. Cinema has long practised the construction of fake narratives, to the extent that many people did not believe in the images of moon walking by the American astronauts, thinking that these images were made in some studio on earth. Despite these doubts, the photographic image was mostly treated as a highly reliable account of the world.

The marketing industry relies heavily on imagery. By juxtaposing different images, the viewer is lured into an attractive atmosphere that is associated with the consumption of a product. In consumer products, the images are largely made through scripted and directed performances. In the tourism and city marketing, there is a large amount of reliance on the selective images of a place, so that it can be shown in a good light, with attractive features and the promise of a good experience. There is no expectation of authenticity from the marketing images, but we allow ourselves to be attracted by the images that we see. We know that the image is manipulated to sell a product, but we are socialized into accepting them as illustrations of reality.

Now, with the invention of images by Artificial Intelligence (AI), and the deepfake technologies that can make and transform images to an unprecedented level, we have entered an age in which images are no longer automatically taken to show reality (Baudrillard, 1995). With the rapid development of this technology, distinguishing the true from false imagery becomes ever more difficult. Photos and moving images are now in the same position as painting and text have been long before. Since the earliest paintings in the caves, images have been a representation of the world that interpreted it without being an accurate copy of it. The depiction of gods in the ancient world was always a representation of an imagined perfection. Between the Renaissance and the nineteenth century, some painters tried to be accurate in their depiction of people, places, and events. But even this was always coloured with an air of fictionality, embellishing the scene and showing the sitter better than they were in real life. Nevertheless, this claim to truthfulness, and its utility as an accurate account of the world, were shattered by the invention of photography.

Writing is a form of image-making. It is an early form of depicting narrative that has long lost the authority of being taken for granted account of truth, if it ever had that authority. When faced with a piece of writing, it is possible to enjoy it for what it is, a piece of writing, as in poetry and fiction. But there is no guarantee that this visually recorded form of speech has an accurate correspondence to real events. Its authenticity and truthfulness will be judged by other standards. So it may be a fictional narrative, but considered to be authentic not for its accuracy of reporting, but for the qualities of the text itself, such as its relationship to the author's subjectivity, the skills with which the narrative has been constructed, and the strength of communication in the depiction of a human condition.

To judge the authenticity of a written account, we primarily look for the credibility of the author, the venue, and the publisher. We do not automatically believe in anything that has been written and published, but we have learned to be cautious about the writings that we come across. This same treatment will now have to be made about the images that we see. Seeing is no longer believing, but the first step in evaluating and judging. Like writing, it would lead to more 'street wisdom' about the cacophony of words and images that we come across, expected to develop the skills of separating truth from fake, despite the complexity and difficulty of doing so. Like in-person communication, in which we need to develop the skills of evaluating what we hear from the person who is sitting in front of us, recorded words and images would need to be approached with the same level of diligence. These are lifelong skills that we start to develop from our childhood, and almost never complete the skill, as everyone is likely to err in their judgements at some point, trust someone unduly, or reject others with a mistaken hunch.

## Challenges of authentication

AI and its likely impacts on different aspects of life have led to debates and controversies. Earlier concerns about its takeover of control were dismissed that it only performs the tasks that are set for it (Midgley, 2018). But deep concerns for its potential impact on human society have been increasingly voiced by many, including some of its initiators and developers. As Michael Sandel argues, AI has created three major areas of ethical concern: privacy and surveillance, bias and discrimination, and undermining the role of human judgement (Pazzanese, 2020). Widespread use of AI may lead to large-scale job losses, intrusion of privacy, lack of transparency and bias in decision-making built into the algorithms, social and political manipulation, and reproduction and intensification of social inequality. According to UNESCO (2023), 'AI-systems deliver biased results', as the search-engine technology processes large data sets, 'prioritises results with the most clicks relying both on user preferences and location' and 'can become an echo chamber that upholds biases of the real world and further entrenches these prejudices and stereotypes online'.

In the context of our investigation into authenticity, the problem of authentication becomes a topic of concern. In 2016, AI produced a Rembrandt painting 351 years after the death of the painter. A total of 346 Rembrandt paintings were scanned and fed into the self-learning algorithms of a computer, which then produced, through 3D printing, a new painting named 'The Next Rembrandt', which could hardly be distinguished in its texture and technique from the old master's works. The question was, 'But who can be designated as the author? The company which orchestrated the project, the engineers, the algorithm, or... Rembrandt himself?' (UNESCO, 2023). According to UNESCO (2023), 'While AI is a powerful tool for creation, it raises important questions about the future of art, the rights and remuneration of artists and the integrity of the creative value chain'.

AI can take over the role of the author, the painter, the actor, the scientist, the bureaucrat, and many others. It can write essays in the style of a 12-year-old, make paintings in the style of Vincent van Gogh, play the role of old actors, perform the songs of dead singers, find recognizable patterns in very large datasets, write office reports, and perform many other tasks. But it is always difficult to know whether these products are accurate and reliable, and what sources have been used to produce them. The AI may search all available online databases and resources, producing summaries without regard for the rights of the original authors, and without the ability or sensibility to ensure the accuracy of the product or its potential impact on society. In short, it produces content, in apparently reasonable form, without any guarantees for reliability or originality. Any sense of origin may be lost in the complexity and impersonality of the technological process. It magnifies the concerns expressed by Walter Benjamin (1969) almost a century ago.

A related problem about falsification can be found in visual representation. It is possible to generate images of individuals performing tasks and being in places without their presence and involvement. You could see the videos of yourself singing and dancing without ever having done so. In one sense, the generation of fake imagery is seen as entertaining and innovative, showing the capacities of new digital technologies for radical transformation of the processes and tools of representation. In another sense, it shows the possibility of fakery that is embedded in these technologies, and the loss of trust in what you see. You can assert that the video of your singing is fake, but you can never do that for a report of a conflict in a faraway land. When you see sci-fi movies with fantastical creatures, you can easily see that they are not found in the real world. But when something has been made in the image of a real event, it is difficult to separate reality from fiction.

When the word fake news started to be widely used around the issues related to the presidency of Donald Trump in 2016, the accuracy of the more established forms of reporting was being questioned. The polarized politicians accused the opposition of lying, while the news reports and analyses from independent sources, or those that were close to the opposition, were labelled as fake news. As a result, trust in the politicians and in the traditional print and broadcast media dwindled. Facts could be easily denied by the most powerful person in the world, thereby appearing to replace the right by might. The emerging information gaps were filled by personal accounts, broadcast on digital platforms and social media, which could be difficult to authenticate and verify. With the decline in trust in the traditional media and the political establishment, populism grew in politics, and echo chambers and radicalized polarization in the culture. You would only believe the reports that come from your tribe, as you tend to trust them, without the need for any other source. The authority of the tribe authorized the production of alternative facts, overruling the authority of truthful accounting of events.

Scepticism about the news and the accounts of events is not an unreasonable attitude, as any news reporting reflects a series of presuppositions, and is controlled by powerful establishments that have many more motives than simple accuracy in reporting. As such, it is difficult to find purely disinterested news. Editorial decisions about the significance of reports are a major gatekeeping exercise, which leads to emphasizing some events and remaining silent about others. The way reports are written and presented can also be so framed that their factual content is given an alternative interpretation. Moreover, different politicians may make completely opposing interpretations of the same facts and figures. When differences of opinion become irreconcilable through polarization of positions, these gaps in trust emerge. When technology and politics come together in this way, the reliability and truthfulness of information about the world are severely undermined. Images and accounts cannot be authenticated to be a truthful reflection of events.

The importance of trust in the news is the role that it plays in the development and maintenance of democratic politics. A century ago, the American journalist Walter Lippmann (1993 [1927]:4–5), questioned the relationship between the private citizen and the news: the private citizen is 'tantalized too often by the foam of events', as 'He lives in a world which he cannot see, does not understand and is unable to direct'. Private citizens are not able to be informed about everything, and they are not particularly interested either. Accordingly, Lippmann questioned what he saw as the myth of democracy, in which the public is sovereign. Lippmann, who is credited with being one of the pioneers of neoliberalism (Reinhoudt and Audier, 2018), ends up putting his faith in the hands of the elites and expects the public to be merely supporting one of the alternatives that these elites provide. In a well-known dispute, Dewey (1946 [1927]) challenges this level of scepticism, believing in the importance of knowledge and in the agency of the public in managing its own affairs in democratic politics.

These debates about the role of technology and knowledge indicate the challenges that lie in the analysis of authenticity as a relationship between something and an origin, both in terms of the identity of that origin and the quality of the relationship with that origin. If that origin is not a single identifiable author, object, or event, but a series of complicated technological processes, with or without human control, where can authenticity lie, and how can anything be authenticated? The origin of authenticity may either be lost in the mist of time, or be a web of social relations, rather than single identifiable authors or events. It is also possible to observe that the accounts of these events may be fake, unreliable, or merely a surface representation. But as social and technological complexities make it increasingly difficult to find that source of originality, the desire and appetite to do so grows and expands. It is in this desire for truthfulness that the search for authenticity lies.

## Between representation and experience

If authenticity is a truthful relation to a source, how should that truthfulness be understood? The subject of truth, whether it exists, and how to access it, has been a longstanding topic of concern and controversy. For our purpose, in relation to authenticity, at least two types of relationship between a phenomenon and an origin may be found: representational and experiential. This duo roughly corresponds to the third-person and first-person perspectives, and the event-law relations that we saw earlier in this chapter.

In the representational relationship, the theory of truth that is usually adopted tends to be propositional, that is, the test of authenticity requires making a statement asserting that something is truly related to the specified origin, and therefore, it is authentic. In other words, it shows a process of authentication. The criterion that is used in this theory is correspondence, that is, the statement that describes something corresponds accurately to that thing. In a narrow sense, it means 'truth is correspondence to, or with, a fact', and in a broader sense, it refers to 'any view explicitly embracing the idea that truth consists in a relation to reality' (David, 2022). There are a number of competitors to this theory, such as coherence theory, which holds that truth lies in the internal relations of beliefs, and pragmatist theory of truth, which is about what works, and a deflationary theory, which holds that truth does not exist. However, the correspondence theory of truth is the most common theory that is used in everyday life as well as in scientific discourse. It has been argued that it is a basic attitude of mind, and therefore, its use is essential, even in fields such as psychoanalysis (Hanly, 1990).

This could be called epistemic authenticity, where authenticity refers to a form of knowledge that is tested through the accuracy of information and representation. In most cases that we have discussed, in reference to the author, object, and event as the points of origin and reference, the process of authentication requires establishing this correspondence. Therefore, if someone claims to be exhibiting a painting by Van Gogh, the truth of this claim needs to be established by scientific methods, historical documentation, expert opinion, and other methods of verification. The result is presented as an objective statement, confirming that this painting is truly the work of the master. A correspondence has been established between the claim and the origin, which in this case is the named artist, the author of the work; thereby the work is accepted to be authentic. This theory is also widely used in the critique of digital culture and the problem of fake news and fake images, where the correspondence between the representation and facts has been doubted, thereby undermining the authenticity of the visual or verbal accounts. Some such accounts may not be authenticated, as there are no easy ways of examining their accuracy.

Another relationship between phenomenon and origin is experiential and intuitive. It draws on the concept of truth as disclosure, which was a key concept for Heidegger, who argued against the propositional and correspondence theory of truth (Heidegger, 1962). For Heidegger, the ground of truth is not assertion, as in the correspondence theory, but in the disclosure of the self. For him, the point of reference was the self, which he called Dasein, and finding truth required going through a temporal process of opening. Authentic existence is the result of relying on the self and its potential to unfold and be discovered. Society and the public were the sources of inauthenticity, against which the self could assert itself to achieve authentic existence. This intuitive disclosure is experienced by the self, but it is not necessarily visible to others, thereby limiting authenticity to a level of unfolding openness as felt and interpreted by the person. However, this remains a first-person determination, which could not be easily authenticated by others. Nevertheless, the social nature of the self does not lead to a detached sense of authenticity, but one that is embedded in its context and mediated through the recognition by others, which unknowingly may have been internalized. This form of truth may be called artistic truth, which is based on a process of imaginative disclosure (Zuidervaart, 2004).

The tension between correspondence and disclosure, between accounting and experiencing authenticity, between epistemic and experiential authenticity, can be found in a historical transition from sincerity to authenticity. At an earlier period of history, the concept of truthfulness was expressed as sincerity, and it was only later, primarily in the twentieth century, that it was replaced with authenticity (Trilling, 1972). The word sincerity, which entered the English language in the sixteenth century, coincided with the emergence of a sense of individuality, and how the public person could be different from the private one. The growth of this discrepancy between public appearance and personal feeling, which led to the development of new social manners (Elias, 1978; Thomas, 2020), also led to the development of literature that aimed at revealing the hypocrisy of public appearance. Trilling (1972:61) argues that 'The literature to which we give our admiration and gratitude fulfils its function exactly by rending the false veil of politeness, by refusing the compromises of urbanity'. Sincerity was a demand for truthfulness.

People were ethically expected to be sincere but socially they were demanded to be polite, causing a contradiction that could not be easily solved, if at all. Trilling's equating urbanity and society with compromise, and the art as an alternative space in which these compromises can be put aside, at least in imagination, shows a conviction about the underlying function of art, and of the concept of authenticity. Living in society, which was once taken for granted, becomes a challenge as urban populations grew and neighbours became strangers. With the emergence of the concept of authenticity, both art and life are expected to be truthful, devoid of the hypocrisies and compromises of urban life. Authentic art allows articulating a sphere of life and thought in which these restrictions are dissolved or wished away. For Trilling (1972:94), 'authenticity is implicitly a polemical concept, fulfilling its nature by dealing aggressively with received and habitual opinion, aesthetic opinion in the first instance, social and political opinion in the next'.

The representational concept of authenticity is widely used in the practices of authentication, as we shall see in the discussion on cultural heritage in Chapter 6, confirming the relationship between a phenomenon and a point of reference through providing some evidence. The social nature of this evidence would make the judgement of authenticity context-dependent and even provisional. But nevertheless, it is a judgement that others can accept or refuse. The intuitive and experiential concept of authenticity is one that is based on the notion of disclosure, which is not accessible to others, and is primarily based on a self that refuses social conventions. The social nature of the self makes it impossible for this self to be detached and in a void. But again, it is a judgement that individuals make for themselves, and only they can be convinced of this authenticity, as the judgement of others can only be a limited one. To go beyond the dichotomy of representation and intuition, Adorno argued for the social basis of authenticity, as we saw earlier (Adorno, 2002; 2004).

In some sense, both concepts of representational and experiential truth are attempts at establishing a direct relationship between an origin and an expression. In the case of the representational truth, the origin is an event in the world, and authenticity becomes a relationship between an outside event and an account. In the case of experiential truth, the origin lies within the body, in its temporality and potentiality, and authenticity becomes an expression of that inner experience.

## Authenticity and event

In the city as a process of unfolding events, each event may be selected as a point of reference, but these events are always in relationship with patterns and regularities, thereby singularities and generalities cannot be understood without each other. From urban crowds to transient images, the urban space is a place of ephemeral events, but these events all have a trajectory, as part of a chain of events. The search for authenticity cannot select one and ignore the other, as the random events and the regular orders are intertwined in urban life.

Recurring events may turn into habits, which are a substantial part of urban processes, whereby people follow everyday routines, within structures that are set and reproduced all the time. Much of the discourse of authenticity takes place in opposition to this habitual city, where routines and habits are seen as inauthentic, determined by others, turning the individual into a follower of conventions and habits, a cog in a system that denies their individuality and blocks their path to recover their originality. The necessity of living together in increasingly complex society, which has been extended and further complicated with the advent of new technologies, creates a discrepancy between what is presented and what is felt.

Habits may become second nature for people, through experience, education, and enforcement, and this second nature has been advocated to be the only place in which ethical behaviour is possible to develop. Nevertheless, the discourse of authenticity looks for originality in chance events and unique occurrences, where habits and conventions are challenged and changed. While some have privileged large-scale, substantial events as the primary source of truthfulness, all events, large and small, can be significant for some individual or group, and as such these minor events cannot be dismissed as unimportant. The events may be fleeting images, individual performances, and individual courses of life. Nevertheless, no event takes place in a void, and even if an event is adopted as the point of reference for authenticity and truthfulness, it is not a detached singularity, but a piece in a larger composition of related events. The point of origin, which may be an outside event or an inside experience, is linked to an expression, an account, that is used to authenticate an object, an event, or a life story. Self-authentication, which is not accessible to others, is an intuitive process, but social authentication requires discursive representations so that some evidence is produced for demonstrating the relationship between the object and event. The event finds meaning not on its own but in relation to other events.

# 6 Time

This chapter examines the temporal dimension of authenticity, analysing it as an expression of a desire for (re)connection to a real or imagined origin in the past. A particular moment or period in the past is seen as the temporal origin of something, and authenticity becomes a quest or a claim for being linked with that origin, overcoming the ruptures between then and now. In the context of accelerated temporality, whereby the world changes at ever higher speeds, and the lines of continuity with the past are broken or threatened, this search for an origin in the past becomes more magnified. In urban space, this is expressed in the concern for heritage preservation and conservation, in tourism, and in a competition between different periods of history, peeling the layers of the palimpsest to reach the authenticity of an origin. In urban society, the nostalgia for the historical community and the efforts for returning to some semblance of the bonds of community are some of the attempts at authenticity. At the same time, authenticity may also be seen as a pathway to the future, as a potentiality that may unfold in time, whereby it is the present moment that may become the point of origin for the future, even if nothing starts ex-nihilo. While pinpointing a moment as an origin may appear as a rupture in the flow of time, it would also reflect a desire for continuity.

## Looking for a point in the past

If we search for a point in time, what do we find? What does it mean to select a point in time? Is time a collection of events, and if so, how can events in the past be reached with any degree of certainty? Is time a substance that can be captured and measured, an experience that is felt, or a relation that is described? What are the ways in which time can be conceptualized and in what ways do these concepts affect the notion of authenticity?

The ancient philosopher Heraclitus interpreted the essence of the world as change. As quoted by Socrates, Heraclitus had suggested that 'everything gives way and nothing stands fast' and accordingly, 'you cannot step into the same river twice' (Plato, 1997:120, *Cratylus*, 402a). According to this view, it would be hard to find any fixed point in the continuous flow of the world, and a search for an origin as the basis of authenticity would only lead to a

DOI: 10.4324/9781003451778-6

constant flow of ever-changing things. Authenticity would lose its meaning, as there are no origins to be found unless the origins are found in change and transformation, rather than fixity and permanence. Socrates believes that Heraclitus is 'spouting some ancient bits of wisdom that Homer also tells us' (402a), but Socrates is not convinced by Heraclitus and presents an interpretation of Homer's poetry that suggests that flows ultimately originate from a spring.

In a Heraclitan world of flows without fixed points, Socrates argues, knowledge is not possible, as 'no one could know anything and nothing could be known either' (Plato, 1997, *Cratylus*, 440b). As he puts it,

> if there is always that which knows and that which is known, if there are such things as the beautiful, the good, and each one of the things that are, it doesn't appear to me that these things can be at all like flowings or motions.
>
> (440b)

This essentialist interpretation is in line with Plato's division of the world into sensible and intelligible realms, whereby the sensible may appear to be flowing and transitory, but the reality lies in the intelligible, fixed world of origins, to which it is possible to return as a source of authenticity. These origins lie in a timeless universe, unaffected by the Heraclitan flows of change, providing solid foundations for knowledge. Authenticity of something can be ascertained by showing a direct relationship with its Idea, that is, its origin in the higher realm of intelligibility, thereby establishing its truthfulness.

From this perspective, authenticity is made possible by finding a point of origin that is not located in the temporality of the world, but in the timeless atemporality of a higher realm. This interplay of origin and flow, of fixity and change, of reality and appearance, shows how the authenticity and inauthenticity of things are judged. Authenticity requires a reference to that point of origin, fixity, and reality, and anything that is flowing, changing, and appearing is merely transitory and unfit for the claim to authenticity. It is a dialectic of the inside and outside, of an inner truth and an outer appearance, which earlier we have seen to be a fundamental element in the analysis of authenticity.

A different, but related, interplay may be found between two concepts of time that describe it as a substance and as a relation. Plato conceptualized time as an independent, absolute substance, which was created with the universe. In contrast, Aristotle argued that time is a relationship between things, rather than a thing itself (Markosian, 2014). For Aristotle, time was closely associated with motion and change, and it was indeed the measure of change. It did not exist in itself, as it was not possible to capture it: the past 'has happened and does not exist', the future 'does not yet exist', and the now, 'which is not a part of time', and which links and separates the past from the future, 'is not easy to see whether it always stays the same or whether it is always different' (Aristotle, 1991, *Physics*, Book IV, 10). In some ways, Aristotle seems to offer a synthesis of Plato and Heraclitus, a halfway house between fixity and flow, whereby things are in a flow but in doing so they constitute fixed forms.

The absolute view of time was echoed in the early modern period and the rise of new philosophical and scientific worldviews. This concept of time was now influenced by the medieval invention of mechanical clocks and the Copernican revolution that removed the earth from the centre of the world. For René Descartes, time was 'a generalized abstraction from the "durations" of particular bodies' (Slowik, 2014). For Isaac Newton, time was an absolute and independent substance: 'Absolute, true and mathematical time, of itself, and from its own nature flows equably without regard to anything external, and by another name is called duration' (Newton, *Principia*, quoted in Whitrow, 1972:100). Newton's views of time and space dominated the modern sciences until the early twentieth century.

The relational view of time was championed by Leibniz, as formulated in his famous debate with Newton about the ideas of time and space. For Leibniz (1979:89), time was 'an order of successions', rather than an abstract, independent substance. Kant disagreed with both, considering time as an inherent aspect of subjectivity: 'Time is not something objective. It is neither substance nor accident nor relation, but a subjective condition, necessary owing to the nature of the human mind' (Kant, quoted in Coveney and Highfield, 1991:28). With Einstein, however, the relative idea of time becomes fully established, whereby time becomes observer-dependent, and its experience and measurement are related to the position and movement of the observer. As Whitrow (1972:116) writes, 'whereas for Newton time was independent of the universe and for Leibniz it was an aspect of the universe, Einstein's theory leads us to regard it as an aspect of the relationship between the universe and the observer'. Moreover, time was integrated with space in a four-dimensional spacetime, a concept that Minkowski had introduced in 1908: 'No-one has yet observed a place except at a time, nor yet a time except at a place' (Minkowski, quoted in Gray, 1989:186).

Therefore, looking for a reference point in time takes us to a pattern of spacetime events and the relationships between them, as it is no longer tenable to see time as a uniform substance. We have the choice, as Einstein had suggested and Thorne (2013) reiterated, as we saw in the last chapter, of looking for that point in two simultaneous ways. We can adopt the third-person view and find this point of reference as a mathematical point in a four-dimensional universe, or we can take the experiential, first-person view and look into the labyrinth of our memories and experiences to find that point of reference. However, neither option would be sufficient for articulating the point of origin that would serve as a source for authenticity. In the four-dimensional mathematical model of the universe, a point of origin may be identified through its coordinates, but this would be devoid of cultural meaning, which is a primary feature of authenticity. In the personal experience of life trajectory, the past, present and future are all squeezed into the selected moment of the experience, but this point of origin is a personal one, turning any notion of origin for authenticity into a mental event inaccessible to others. It is a personal selection, and judgement, about experiences, habits, and memories, from within a fragmented and forgetful self, with all the limitations that this entails.

Authenticity, however, is always socially mediated and a component of social recognition. For something to be authentic, a judgement is required, which is a social act based on cultural habits, social conventions, and formal institutions. Therefore, we may alternatively look for the cultural frameworks that define a point in time as significant, and critically investigate those frameworks. The notion of public time, in which the presentation and measurement of time are developed through institutions, conventions, and objects, is one of these frameworks. Public time includes clock time, which sets similar frameworks across different places and is universalized through institutions and conventions. Public time also includes historical time, defined by major social events and widely accepted narratives. Therefore, the critical question is about how, and by whom, these social events are established as significant, and how the narratives of history are recorded. When something is dated back to the time of a ruler or the advent of a natural disaster, the range of events in a period of history may be reduced to a single descriptor, which is then picked as the temporal point of reference. The power relations that are embodied by these choices require to be critically unpacked and assessed.

## Interrupted continuities

In addition to the problem of finding an origin in the past, the search for authenticity indicates the need for showing a degree of continuity between that origin and the present. This is one of the places in which the discourses of authenticity and identity overlap. The question of personal identity as continuity over time has been a classical philosophical problem. The question that has arisen is whether a person's identity persists over time through physical or psychological continuity (Olson, 2023). In searching for authenticity in cities, this question of identity may be equally raised: over the course of years, decades, and centuries, does a city remain the same, or does it change to the extent that it no longer has the same identity? Is this continuity to be found in its physical elements, social institutions, habits and behaviours, or memories and feelings of its citizens? Or is this continuity found in the narratives that link these different pieces together to create a coherent story? (Madanipour, 2013).

The question of identity, therefore, asks whether something has an essence, remains the same over time, or always changes. Authenticity is a form of identity: if you describe something as authentic or inauthentic, you have narrated an element of its identity. It is assigning an adjective to something, and in doing so contributing to the creation of that adjective. Narrative identities are made of a series of names and descriptions, which may include authenticity and inauthenticity. If things are multiple and always changing, what is possible to find are shifting patterns, rather than fixed essences that can secure a stable identity. Authenticity is always a quest for temporal coherence in the middle of these shifting patterns, always harking back to a previous period in which things were apparently less corrupted, more truthful, and more authentic.

Narrative identity, according to Paul Ricoeur (1991:77), is created when 'the narrative constructs the durable character of an individual', linking different life events to a sense of continuity. Ricoeur suggests that many of the difficulties about the concept of personal identity can be addressed by distinguishing its two separate meanings: 'identity as sameness' and 'identity as self', arguing that self-ness is not sameness. In analysing this narrative identity, Ricoeur (1991:73) suggests that 'knowledge of the self is an interpretation'. The self, he argues, 'does not know itself immediately, but only indirectly, through the detour of cultural signs of all sorts' (p.80). Therefore, narrative identity is a mediation like other signs and symbols, which 'borrows from history as much as fiction'; the result is that the life story becomes 'a fictive history', 'an historical fiction', like the biographies of great characters which are a blend of history and fiction (p.73). The mediation of this narrative through cultural sources shows how identity as self-recognition is intertwined with identity as mutual recognition, which Ricoeur (2005:21) sees as 'indeed our most authentic identity, the one that makes us who we are, that demands to be recognized'.

In a globalized world, in response to the intensifying patterns of interdependence across different regions and countries, and the forces of cultural assimilation that push to eradicate historical differences, the desire for distinctive and authentic identities has become stronger than ever before. What was taken for granted to be integral elements of being true to someone's culture and history now feels to be threatened by the whirlwinds of global change. In Marx's famous phrase, all that is solid melts into air. In this context, the search for stability and permanence, for consistency in a changing world, becomes stronger. Authenticity, as a reference to some origin, is an expression of that desire to seek stability.

Although the problems of personal identity may show similarities with urban identity, there are significant differences between them. While they are both made of diverse component parts, the human body encapsulates an empirical unity that remains the same over time; even if its parts age, evolve, and transform, they almost always remain bound together. A city is also made of many parts, but these parts do not necessarily work together and may age, evolve, and transform according to different rhythms and rules. While Hegel envisaged individuals and societies to be all driven by an overall spirit, the city in fact runs according to many different beats, the remainders of different periods of history, thereby it tends to be a dissonant collage rather than a coherent whole (Madanipour, 2017). While some of these different parts grow old and become obsolete, others are renewed, finding a new lease of life. While the individual's narrative identity is told by a single storyteller, who may well change those stories frequently and reflect different selves that inhabit the same body, the city's narrative identity is told by many such storytellers, each from a different life experience, social and economic condition, and epistemic perspective, and thereby multiplying the stories by a discordant cacophony of voices. While there are many social platforms that frame these different trajectories, these platforms cannot create a single, continuous, and coherent story, at least in the minds of the urban dwellers.

In addition to this problem of unity and multiplicity, there is also the problem of mind and body dualism, which in the city becomes a problem of social and physical dualism. The philosophical dualism of mind and body, which was championed by Descartes and refused by later generations, has found its parallel in the separation of the social and physical aspects of the city. This separation is clearly visible in the specialization and fragmentation of academic and professional disciplines: those in physical sciences may pay little attention to the social and human aspects of the world, while those in the humanities and social sciences may mainly focus on meanings and relations, without much attention to the materiality that is integral to them. However, a city is made of both physical and social elements, which cannot be separated as independent variables. When a city is physically destroyed by a natural disaster or an urban redevelopment scheme, and its population is moved to a new site, can it be called a new city with a new identity? Similarly, if the people of a city are forcefully moved out and replaced by another population, can it be called the same city? These are interruptions in continuity, and in each case the city is a combination of both physical and social dimensions.

These sudden and dramatic changes of an entire population, or total physical destruction are relatively rare. What is far more common is a gradual change of the different component parts of the city. The lines of temporal continuity are not suddenly broken, but they are stretched and transformed gradually and in a piecemeal manner. Every city experiences births, deaths, new arrivals and departures, decline and renewal of roads and buildings, and change of functions and institutions, which step by step change the city in all of its dimensions. With the rise of industrial cities since the early nineteenth century, the pace of these changes accelerated, resulting in agglomerations that raised fear and concern about social coherence and stability (Briggs, 1968). By our time, when the new century witnessed the population of the world becoming mainly urban, the pace and scale of change had become unprecedented. These changes have been supported by accelerated mobility and intensified communication and their associated transport and digital technologies. Vast urban areas have been formed in a relatively brief period of time, as exemplified in the transformation of Shenzhen in China or Dubai in the United Arab Emirates from a small town to a major metropolis in less than four decades. The lines of continuity with the past are broken or blurred, and the problem of accounting for identity and authenticity is more challenging than ever before.

In this maelstrom, memories are interrupted, making it difficult to construct narratives of continuity. Some have argued that these lines need to be maintained through the materiality of the built environment, through a public realm that connects different generations together. As Hannah Arendt (1958:9) wrote, '[H]uman existence is conditioned existence, it would be impossible without things, and things would be a heap of unrelated articles, a non-world, if they were not the conditioners of human existence'. For her, this public realm could guarantee continuity and permanence:

> Only the existence of a public realm and the world's subsequent transformation into a community of things which gathers men together and relates them to each other depends primarily on permanence. If the world is to contain a public space, it cannot be erected for one generation and planned for the living only; it must transcend the life-span of mortal men.
>
> (Arendt, 1958:55).

*Figure 6.1* In Barbès, the physical environment has remained the same, but what was once a vibrant, North African street market has now been removed from the city by the authorities (Paris, France)

In our time, however, when digital technologies and the ever-faster pace of urban transformation undermine the possibility of such continuity and permanence, the challenge for temporal authenticity is to rethink, reconstruct, and maintain some lines of connection between the present and many different points in the past. A single, fixed point of origin could only be no more than a fictional account, and a more realistic approach would require the inclusion of many such points in the past and in the present, and in their meanings and shifting patterns of relationships. In response to the restless change and interrupted identities, some lines of continuity could be sought in the materiality of a public space. The problem facing diversified and stratified societies, however, is the question: whose public space? (Madanipour, 2010) (Figure 6.1). Continuity may be sought in the heritage of the past, but who decides what counts as heritage, how, and for whom?

## Searching for connections: authenticity in world heritage

Authenticity is a central concept in the world of cultural heritage, where it is used in both describing and evaluating the heritage items, that is, how something is authentic and how its authenticity can be determined. In other words, the term is used in both ontological and epistemological capacities. It was partly based on a conservative paradigm that placed its point of origin in the past, reacting against the modernist paradigm that chose the present as its point of origin for the future.

Authenticity is considered to be 'an essential element in defining, assessing, and monitoring cultural heritage' (UNESCO, 1994), and one of the most influential factors in deciding how to deal with the conservation and restoration of historical inheritance. Buildings, objects, and environments deteriorate over time, and the decisions about what to keep and how to maintain them have varied in different periods and according to diverse paradigms. In buildings and cities, historical layers accumulate over and alongside each other, resulting in an environment that has many origins, histories, meanings, stakeholders, and interests. In archaeology and conservation, authenticity started from the late seventeenth-century idea of 'the material condition of something original and unique', evolving to the twentieth-century notion of authenticity as 'the presence of original material' (Benjamin, 1969; Odegaard and Cassman, 2020:1228). Despite this historical span of attention to the concept and the vast literature that has been produced on the subject, many questions remain.

These questions have been widely discussed by scholars, policymakers, and practitioners. In this brief section, I mainly focus on the material that is produced by the United Nations' cultural arm, UNESCO, on the subject of World Heritage. In producing these documents and guidelines, much effort has been made by international teams of experts drawing on different cultural contexts to articulate the concept of authenticity in cultural heritage.

These guidelines have been widely used by governments, academics, and professionals in the nomination and designation of the heritage that merits global recognition and support.

The manifesto for modernist urbanism, the 1933 Charter of Athens issued by the *Congrès Internationaux d'Architecture Moderne* (CIAM), advocated radical reorganization of cities on a rational, functional basis (Sert, 1944:246–249). It included an article on the conditional preservation of historic districts: 'Buildings or groups of buildings that are remnants of past cultures should not be demolished' (p.248). However, for this preservation, buildings and neighbourhoods needed to meet three conditions: they should be real representatives of their period, and therefore of general and educational value; their presence should not affect the health of their resident populations; and their location should not affect the construction of thoroughfares, the conditions of urban traffic, or the organic growth of the city.

These conditions acknowledged the value of historic buildings and neighbourhoods but subordinated their fate to the functional demands of health, movement, and growth. Adapting the new to the old was not an option: 'All attempts at adapting new districts to these old layouts (which is often done under pretext of preserving local characteristics) have had bad results. Such adaptations to the past should not be tolerated in any case' (Sert, 1944:248). Historic monuments had to be disconnected from the declining areas that surrounded them:

> By a planned clearance of slum areas, which are frequently to be found in the neighbourhood of these monuments of the past, it is possible to improve the living conditions of the residential areas near-by and to safeguard the health of their inhabitants.
>
> (Sert, 1944:248)

This attitude was adopted by city designers and public authorities around the world, with an optimistic view of the future that disregarded the past, an attitude that continues to prevail in many parts of the world. The past was another country, backward, poor, and unhealthy, better left behind, or be replaced by a shiny, healthy, and well-ordered city of the future.

In response to the impact of the modernist attitude, a generation later the Second International Congress of Architects and Technicians of Historic Monuments produced the 1964 Venice Charter, the International Charter for the Conservation and Restoration of Monuments and Sites, which is a landmark document in the protection of cultural heritage. It explicitly mentions the contribution of the Charter of Athens to the development of an international movement, and how it had been followed by documents and initiatives by ICOMOS and UNESCO. Now it was the time to look at the Charter of Athens 'afresh', to examine its principles and 'enlarge its scope' (ICOMOS, 1965). The Venice Charter argues that historic monuments are the living witnesses to people's age-old traditions, and as such they are humanity's common heritage, which demands the responsibility of safeguarding them for future generations: 'It is our duty to hand them on in the full richness of their authenticity' (ICOMOS, 1965). In the context of accelerated temporality and radical transformation, a sense of instability, vulnerability, and anxiety is generated. The quest for authenticity is a quest for stability and secure identities; the Venice Charter was a clear response to the Athens Charter, demanding a pause in the latter's urge for the wholesale redevelopment of cities, and the sense of disconnection and disorientation that it seemed to generate.

In its effort to enlarge the notion of historic monuments, the Charter's first article goes beyond single buildings and includes its urban or rural setting as 'the evidence of a particular civilization, a significant development or an historic event' (ICOMOS, 1965: Article 1). It also goes beyond the great works of art to include the 'more modest works of the past which have acquired cultural significance with the passing of time' (Article 1). These monuments are not just works of art but also historical evidence, and they should be conserved and restored with the help of science and technology (Articles 2 & 3). Conservation would be supported by finding a 'socially useful purpose' for the monument, but this should not change its layout, decoration, or setting (Articles 5 & 6). 'A monument is inseparable from the history to which it bears witness and from the setting in which it is located' (Article 7). Restoration should aim at preserving and revealing 'the aesthetic and historic value of the monument', 'based on respect for original material and authentic documents' (Article 9). Any additional work that may be necessary should be kept distinct, bearing 'a contemporary stamp' (Article 9). However, 'The valid contributions of all periods to the building of a monument must be respected since unity of style is not the aim of a restoration' (Article 11). Removing these layers should only be done in 'exceptional circumstance'.

The Venice Charter implies that authenticity in a monument, broadly defined, means a truthful relationship to an origin in the past, which could be 'a particular civilization, a significant development or an historic event' (ICOMOS, 1965: Article 1). It acknowledges that this history is multi-layered, and therefore cannot be purified by pealing back the layers to reach an original point in time. It also implies the necessity of maintaining the integrity of the different parts of a monument: building, layout, decoration, and setting. It therefore acknowledges the temporal and spatial layers that together make a monument.

However, after this Charter, a need was still felt for further articulation of the concept of authenticity and its implications for cultural heritage in different contexts. In particular, the Japanese experts expressed concerns about cultural differences, where the practices of conservation and restoration were widely different from their Western counterparts (Stovel, 2008) (Figure 6.2). This need led to the development of the 'Nara document on Authenticity' by a group of international experts meeting in Nara, Japan, in 1994. Their task

*Figure 6.2* Cultural differences are expressed in different temporalities, and in different approaches to the heritage of the past (Sendai, Japan)

was to define and assess authenticity, especially as it was noted that in some languages there is no word for this concept (UNESCO, 1994). This ocument continues the process of enlarging the scope of cultural heritage that the Venice Charter had started. The Venice Charter expanded time and space by going beyond single buildings and single histories to include the settings and subsequent histories. The Nara document now expanded it further to include the diversity of cultures and forms of expression. This was a shift from the Venice Charter's belief in universal absolutes to a more relative and contextual understanding of judgements in conservation (Stovel, 2008).

In its effort to articulate the concept of authenticity, the Nara document refers to three levels of significance: values, attributes, and information. First, the selection of something as heritage is embedded in the cultural values of a particular society. 'Conservation of cultural heritage in all its forms and historical periods is rooted in the values attributed to the heritage' (UNESCO, 1994, Para 9). When cultural properties are proposed for the World Heritage list, they are examined for their 'outstanding universal value', which is tested with the yardstick of authenticity, so it is important that this is done with 'full respect to the social and cultural values of all societies' (UNESCO, 1994, Para 2). As UNESCO (2017, Para 49) explains, 'Outstanding Universal Value means cultural and/or natural significance which is so exceptional as to transcend national boundaries and to be of common importance for present and future generations of all humanity'. The potential tension between the universal and the particular, as shown in the universality of outstanding cultural values that is the hallmark of the World Heritage list and the particularity of cultures and forms of expression, is addressed through an inclusive and practical approach, as 'the cultural heritage of each is the cultural heritage of all' (UNESCO, 1994, Para 8).

Second, the cultural property under scrutiny is analysed through its attributes (UNESCO, 2017, Para 82). The Nara Document acknowledges the diversity of forms of cultural expression. The need for acknowledging the diversity of cultures is directly related to acknowledging the diversity of their forms of expression, going beyond the traditional emphasis on monuments and sites to include cultural landscapes and intangible heritage. 'All cultures and societies are rooted in the particular forms and means of tangible and intangible expression which constitute their heritage, and these should be respected' (UNESCO, 1994, Para 7). These attributes are listed as 'form and design, materials and substance, use and function, traditions and techniques, location and setting, and spirit and feeling, and other internal and external factors' (Para 13). These attributes have remained almost the same since the publication of the document but now expanded to include 'management systems' as well as 'language and other forms of intangible heritage' (UNESCO, 2024; 2017, Para 82). The listed attributes express cultural values, so authenticity, which 'only applies to cultural properties and to the cultural aspects of 'mixed' properties', is 'the link between attributes and Outstanding Universal Value' (UNESCO, 2024). In other words, authenticity is 'the essential qualifying factor concerning values' (UNESCO, 1994, Para 10).

Third, to make the link between attributes and values, the test of authenticity relies on the sources of information that need to be 'credible or truthful' (UNESCO, 1994, Para 9). Therefore, it states that 'the essential contribution made by the consideration of authenticity in conservation practice is to clarify and illuminate the collective memory of humanity' (Para 4). The sources of information are 'all physical, written, oral, and figurative sources which make it possible to know the nature, specificities, meaning, and history of the cultural heritage' (UNESCO, 1994, Definitions). In other words, these sources, if reliable, can prove the existence of a truthful relation between a cultural property today with a point of reference in time and culture. Authenticity implies originality in the cultural property itself, which can only be affirmed through the reliability of the sources of information that prove that originality. It engages in authentication, establishing a believable correspondence between the sources of information and the item that expresses cultural heritage. However, across and within cultures, there are differences in the credibility of the sources of information and in the values given to cultural properties. Therefore, fixed criteria cannot be used for the judgement of values and authenticity, and these judgements must be made within their own cultural context (UNESCO, 1994, Para 11).

Nevertheless, many questions remain about the practices of heritage conservation and restoration. For example, the Riga Charter on authenticity, which addressed the concerns of the newly independent states in Eastern Europe, declares that 'replication of cultural heritage is in general a misrepresentation of evidence of the past, and that each architectural work should reflect the time of its own creation' (ICCROM, 2000, Para 5). Socialist modernism in these countries had demolished many old buildings in the name of modernization, and now there were some attempts to rebuild them in their original form. In Warsaw after the destructions of the Second World War, the city had been rebuilt in the original form, as an expression of resistance and assertion of national identity. A similar attitude was now being expressed in the post-Soviet lands, as an attempt to delete the traumatic layers of history and recreate pleasant memories of the past. In many Eastern European cities, treatment of the layers that Nazism and socialism had added to cities now became a serious challenge. The question, therefore, is whether such reconstructed forms of signification can be accepted as an authentic expression of a culture, and what directions should be adopted when many contested layers constitute the history of a city.

This leads to a broader question about the link between a work and a time. A work may be factually shown to be built at a particular point in time, but it cannot be taken to reflect its time. Such works may reflect the dominance of a ruler or an ideology, which was not accepted or shared by most of the population. There may be dominant styles of architecture that are taken to reflect a time, but historical periods may not be reduced to single approaches in architecture. What if a work of architecture is anachronistic in relation to this dominant paradigm? The notion of zeitgeist has been used to show what types of activities represent the dominant trend in an era, but this is ultimately a selection, based on stressing some objects and activities, and ignoring others. The Hegelian notion of Spirit is an attempt to give coherence and meaning to periods of history. But any period of history is full of diversities and contradictions, which may not be possible to reduce to a single spirit.

This shows that the relationship between a work and a time can be expressed in two distinctive ways. First, it can be dated to a particular time; second, it can be claimed to 'represent' that time. The first requires reliable evidence to prove that dating, but the second is a value judgement based on how to describe a period of time through a selected set of themes and symbols. As we saw in Chapter 4, any work reflects its time, knowingly or unknowingly, just by virtue of being produced in that period, but this does not mean that it can represent that time.

The Nara document attempts to address the contested notion of universal values, as values that go beyond the national and cultural borders, rather than defining the universal as values that everyone necessarily holds. It somewhat relativizes the claim to universality by arguing that the value of heritage is determined in a particular context, but it can go beyond this context and be appreciated or shared by others. Nevertheless, a claim to universality is always exposed to critique from those perspectives, within and across national borders, that may find themselves excluded or would not align with its framework, or from those who may dispute the processes of evaluation and selection.

The process of applying for listing something as World Heritage requires presenting it to those in authority and proving to them that it is what is claimed to be. This requires a course of judgement by that authority, testing the three levels of values, attributes, and sources of information. The claim is judged according to the values that it has for a cultural community and beyond, the listed attributes, and the credibility and truthfulness of the sources of information on which it depends. It is this judgment that confirms or rejects the originality of something. Authenticity becomes a judgement of the originality of something and of the credibility of our knowledge about it; it is based on scientific and cultural evidence, to confirm or deny the relationship between something and an origin in the past, as expressed through its listed attributes. Inclusion in the World Heritage list requires establishing a temporal link between now and a reference point in a past time.

According to our investigation, this version of authenticity uses time as its point of origin and a correspondence theory of truth to prove that link to the intended origin. Even when current practices are included in the World Heritage list, their value is derived from their historicity and continuity, showing the temporality of the concept in this perspective. It locates the concept of authenticity in the representation of a culture and society, stressing that indicators such as the spirit and feeling of a cultural property are important but not easily accessible for practical assessment (UNESCO, 2017, Para 83). The Nara document, which became the primary effort in defining authenticity in relation to historical heritage, was engaged in practical considerations, avoiding the questions of memory and identity that were raised by some participants (Stovel, 2008). It is to these questions that we now turn.

## Constructed narratives

In selecting a cultural property as a significant heritage item, major questions arise about how to decide what constitutes the original material, how far back should we go in search of that origin, and which of the many origins in a place should be given priority in such decision-making. Further questions are about who is entitled to make that decision, and on what basis; and when it comes to the practices of archaeology and conservation, how do we deal with what we find? More broadly, these questions are about: what signifies the past, which past is selected, whose past is it, why it needs support, who deals with it, and how, and what it means for others.

The quest for authenticity in cultural heritage rests on the sources of information, the documents that can prove the authenticity of a historical item. This requires finding authentic documents that can prove the authenticity of the item itself. However, approaching documents is not a simple process of a historian gathering, reading, and assessing the authenticity and veracity of documents, before writing them up. This has been called 'epistemological naïveté' (Ricoeur, 2006:177), as the historian starts with a question. 'The documents do not speak unless someone asks them to verify, that is, to make true, some hypothesis. Therefore, there is an interdependence among facts, documents, and questions' (Ricoeur, 2006:177). The documents, which can include 'everything [that] can be interrogated by a historian with the idea of finding there some information about the past', are meant to provide facts that can be 'asserted in singular, discrete propositions, most often having to do with the mentioning of dates, places, proper names, verbs that name an action or state' (p.178).

However, facts are not the same as events. Ricoeur (2006:178) warns,

> A vigilant epistemology will guard here against the illusion of believing that what we call a fact coincides with what really happened, or with the living memory of eyewitnesses, as if the facts lay sleeping in the documents until the historians extracted them.

Facts are not the events themselves, but statements that are meant to represent events, and therefore they are constructed narratives, as we saw in Chapter 5. This does not indicate that they are necessarily fictitious, but that they are not free from the presuppositions of the researcher, the questions that are being asked, and the context in which these questions are raised. The sources of information that are used for authenticating the items of cultural heritage are an inseparable part of that cultural item. The narratives that are constructed to describe and authenticate a cultural property are a necessary ingredient of the property's authenticity. As such, they are embedded in epistemological, political, and ethical conditions.

The gap between the present and the past, Gadamer (2004) argues, can be addressed through a 'fusion of horizons'. In his critique of historical consciousness, Gadamer questions the nineteenth century historicism and its attempts to turn history into a modern science. He is critical of applying the methods of natural science to human sciences. He is also critical of the empiricist concept of experience, to say that any experience is imbued with prior ideas, rather than a pure encounter; he argues for a hermeneutic experience, which follows Aristotle in seeing the experience as an active encounter with the world, rather than a passive one in which the observers only receive sense data. Drawing on phenomenology and Heidegger, he argues for a hermeneutic approach, which is based on the Platonic model of dialogue, and partly on the Hegelian dialectics, when the conversation transforms both sides. At the same time, Gadamer tries to show the limits of reflexivity and argues for recognizing the traditions of which we are a part. However, this conservative emphasis on traditions is rejected by critical theorists such as Habermas (Mendelson, 1979). Others have argued for the necessity of a critical approach to hermeneutics: if hermeneutics is a theory of interpretation, it needs to include a critical element. Drawing on Derrida's theory of deconstruction, this critical approach argues for a 'radical hermeneutics', which takes 'the point of view of the outliers, the outsiders, the ones whose views have been neglected or excluded' (Caputo, 2018:10).

Therefore, authenticity is part of a narrative identity, whereby a truthful relation is claimed to a point of reference in the past. This point of reference may be difficult to pinpoint, as it is located in an endlessly receding chain of events. Furthermore, the narrative claims that relationship is always mediated through the cultural context, social institutions, personal memories, and active experiences. The narrative seeks social recognition by bringing together different fragments and perspectives in a line of coherence and continuity; as such it needs to be exposed to critical examination, especially when it comes to the complex world of the city.

In searching for a point of reference in the past, whenever some stable relationships are identified with distinguishable expressions, they are selected as the origins of authenticity. They become nodes in a pattern of contingency and change; it is these points that are picked up as the platforms for originality. This can be seen in the history of a city, when particular periods of prosperity and development leave their mark, and while these marks may be a combination of various ingredients, their composition is sufficiently durable and recognizable to be selected as a point of origin. In returning to an origin, the past is reinvented and relived, rather than revived, which in most cases would be impossible. Therefore, searching for an origin in the past, a tradition, is partly an innovation, a projection of the now to the past, a future-oriented variation on a theme.

Heritage as a narrative is always selective, emphasizing some aspects of a past and ignoring or forgetting others. It has many sides, and some of these sides are celebrated and others remain invisible, all reflecting the unequal power relations at the time and the following times. Walter Benjamin (1969:256) warns that what are called 'cultural treasures' are often the spoils accumulated by rulers and their descendants, and therefore these should be viewed 'with cautious detachment':

> For without exception the cultural treasures he surveys have an origin which he cannot contemplate without horror. They owe their existence not only to the efforts of the great minds and talents who have created them, but also to the anonymous toil of their contemporaries. There is no document of civilization which is not at the same time a document of barbarism.
>
> (Benjamin, 1969:256)

Adorno also draws attention to the dark side of what appears to be a harmonious and attractive legacy of the past. As he argues, 'Contemporary representations of medieval towns usually look as if an execution were just taking place to cheer the populace' (Adorno, 2004:191). Adorno argues that any harmony that may have existed at that time would have been similar to the most recent ones: 'pressure-born and brittle'; as in fact, 'only as lost conditions do they become glamorous' (Adorno, 2004:191). In this glorification of a place, the memories are reconstructed, ordered, and sanitized, while the unsavoury sides are kept in the subconscious memories of the place. An example is a medieval street called Shambles in York, which is a popular part of the city for visitors, with its neat narrow, winding street, and its colourful shops in well-maintained timber-framed buildings. A tourist may see a coherent urban environment in its architecture and an imagined cosy social life. However, this was a place where animals were slaughtered, with blood and carcasses everywhere, very different from what is visible today. The harmony

*Figure 6.3* The experience of visiting the medieval streets does not reflect the conditions of life in the medieval city (Munich, Germany)

that is seen is the result of a detachment between what was and what we experience, between the complex past and the edited present. Across Europe, medieval towns attract millions of tourists every year, keen to experience an atmosphere that feels coherent and harmonious, quite unlike the fragmented world in which they live in their everyday lives (Figure 6.3).

When an object is separated from its context, it may become an object of aesthetic interest, devoid of its cultural meanings, and open to multiple, ambivalent interpretations (Figure 6.4). The object may be placed in a museum, where it sits side by side with other objects from different places and times; even if its history and context are well explained, it merely provides a fleeting experience for most visitors to the museum. It can also be placed in the corner of a street, where all the histories that led to its construction and location may have been lost in the mist of time, becoming an everyday object that no one notices, or one that is visited by tourists who see it as a part of their own enjoyment, rather than what it originally meant, or even what it now means for the local population. The object is reduced to an appearance with little meaning, at most a vehicle of a momentary aesthetic experience.

Another tension in heritage is the differences of value that are assigned to cultural properties, and the gap between rhetoric and practice. While much of the heritage protection is justified and based on the appreciation and enhancement of cultural values, in practice there are many financial considerations at

*Figure 6.4* Displaced objects are detached from a context, and in the new context their meaning is open to multiple interpretations (Lisbon, Portugal)

work, which lead to the commercialization of heritage properties. The close relationship between tourism and heritage provides the necessary resources for heritage protection, but it also exposes heritage areas to the workings of the tourism industry, real estate considerations, and a market perspective that is not necessarily immersed in cultural considerations. Conservation and restoration of heritage properties are rightly justified as a culturally valuable process, but, at the same time, they may be used to stimulate investment and reward, which could narrow down the range of possibilities and limit social access. Tourism-based heritage conservation could help with the regeneration of rundown areas, but it could trigger the gentrification of an area, displacing the existing populations in these areas. When cultural and economic considerations are intertwined, potential fault lines are created between a socially acceptable rhetoric that is on the surface and an economically driven course of action that could undermine the social and cultural aims of the process. It generates an inherent contradiction that may not be easily resolved.

The contested nature of cultural heritage can be observed in many areas of social life. For a long time, the support of some forms of cultural practice has been questioned as elitist. In the UK, for many years questions have been asked about the reasons why opera as an artform needs to be subsidized, based on the argument that it is only interesting and affordable to a small and wealthy part of the population. The clash of social classes, which shaped the political structure of the country, has been reflected in the attitudes towards arts and culture, partly reflected in a confrontation between high culture and popular culture. The middle-class culture and its symbols have frequently been attacked for the neglect and even suppression of the working-class culture. Resisting this neglect included the campaign for the preservation of the industrial heritage, which had been debased through deindustrialization, globalization, and the decimation of the industrial workforce. In fighting suppression, critics have demanded equality of expression for the voiceless (Rancière, 2013), so as to include women, lower-income groups, cultural and physical minorities, and the colonized populations in cultural expression and heritage preservation.

## Contested memories

The fragments of the past may be brought together to construct a coherent narrative, building a bridge between the present and many different points in the past, a bridge that also serves as a platform for the future. These fragments may be brought together as expressions of some universal archetypes made of deep memories (Jung, 2014), which make up a universal culture (ICOMOS, 1965). Collective memories and cultural expressions are intertwined to claim universality and inclusiveness. However, collective memories and cultural expressions are always partial and perspectival, produced by some groups from a specific experience and with a particular level of power. The claim to universality may appear to have been accepted, but their internal inconsistencies remain dormant until a moment of crisis arrives. At that moment, the partiality of memories and expressions becomes revealed. Whose memory, what is selected as a representative of the memory, preserved by whom, and for what purpose? (Figure 6.5).

In 2020, anti-racism protestors in Bristol, UK, tore down the statue of the seventeenth-century slave trader Edward Colston and threw it in the river (BBC, 2020). Colston was a London-based merchant who made a fortune through the slave trade, a fortune which he later moved to money lending. He was a member of the Royal African Company that had the monopoly of West Africa's trade in the late seventeenth century, a company that between 1627 and 1689 sold 100,000 slaves in the Caribbean and the Americas, branding all the slaves with its initials RAC on their chests (Farrer, 2020). He was also a politician, and a philanthropist who made major donations to charities in London and the city of his birth, Bristol. His support for almshouses, schools, hospitals, and churches made him a prominent figure in Bristol, recognized by erecting his statue in 1895 in an avenue that was named after him, and by giving his name to a school, a concert hall, a high-rise office block, and a street. Controversy over his statue had been raging for some years. A petition signed by thousands of people had requested the removal of the statue, declaring that, 'Whilst history shouldn't be forgotten, these people

*Figure 6.5* Inukshuk, an ancient landmark for navigation, and a symbol of hospitality and friendship in the Inuit culture, competes for attention with the towers of the English Bay (Vancouver, Canada)

who benefited from the enslavement of individuals do not deserve the honour of a statue' (Quoted in Farrer, 2020). It was finally in 2020 that the statue was removed by a group of protestors, triggering new controversies about the legality of such a move.

In 2020, anti-racism protestors sprayed the statue of Winston Churchill in London, calling him a racist (BBC, 2020). Churchill was the wartime prime minister who had been selected the man of the twentieth century by the BBC viewers, and in 2002 was voted to be 'the greatest ever Briton' (Heyden, 2015). Without him, a piece in the *Washington Post* at the turn of the millennium stressed, 'the world today would be unrecognizable -- dark, impoverished, tortured' (Krauthammer, 1999:A31). At the same time, his comments to the 1937 Palestine Commission are shocking today:

> I do not admit that the dog in the manger has the final right to the manger, even though he may have lain there for a very long time. I do not admit, for instance, that a great wrong has been done to the Red Indians of America, or the black people of Australia. I do not admit that a wrong has been done to those people by the fact that a stronger race, a higher grade race, or, at any rate, a more worldly-wise race, to put it that way, has come in and taken their place.
>
> (Quoted in Parliament, 2021)

These comments show a social Darwinian view that has triggered many controversies (Heyden, 2015). It implies that it is not to a point in time that we should go back to find an origin, and even if we find such an origin, what matters is that an alternative, new origin is created and justified by might, overruling and replacing the other origin. It seems to associate right with might, a view expressed millennia ago by the sophist teacher Thrasymachus, that 'justice is nothing other than the advantage of the stronger', a view against which Socrates argued tirelessly (Plato, 1997, *Republic*, I:337c).

In 2024, the statue of Captain Cook in Melbourne was toppled on the eve of Australia Day, with graffiti on the statue's empty plinth promising that 'The colony will fall' (Ritchie, 2024a). Australia Day is celebrated by most Australians each year, marking 26 January 1788, when the first British fleet landed in Sydney Cove. For the existing inhabitants, however, this was the beginning of losing their land and culture, hence naming it the 'Invasion Day'. This has led to two competing stories about the day: 'One is of nation-building and achievement; the other is of the displacement and dispossession of Aboriginal and Torres Strait Islander people' (Ritchie, 2024b). The single point in time becomes an origin to which people go back for narrating their history, but this is never a single history.

After independence from the British Empire, many symbols of British rule, including the statue of Queen Victoria, were removed from Indian public spaces. In 1957, prime minister Jawaharlal Nehru proposed a nuanced approach: splitting the statues of the colonial period into three groups and dealing with them accordingly: those that offended the 'national dignity', those that had 'historic significance', and those that were 'artistic' (Ansari, 2020). Removing the symbols of the colonial era in India included street names, and eventually cities' names, as exemplified in Mumbai and Varanasi, and even discussions about changing the country's name from India to Bharat. With the rise of religious nationalism, the removal of the layers of history intensified, going back further for centuries, removing the symbols of the Muslim rule by Mughal emperors over India, seeking to assert a single, Hindu identity for the country. This is exemplified in the highly symbolic demolition of the sixteenth-century Babri mosque in Ayodhya in 1992 (BBC, 2017), to be replaced by a grand Hindu temple, which was opened by the Hindu nationalist prime minister in January 2024.

In 2021, a series of anti-government protests by indigenous Colombians toppled and disfigured the statues of the Spanish conquistadors and even the statue of Christopher Columbus, whose name was given to the country (BBC, 2021). This is another version of dealing with heritage, deleting some layers of the palimpsest, from an emancipatory perspective, a critique of power and inequality, aimed at decolonizing the historical memory. For the conquered people, it would be absurd to celebrate the triumphs of the conquerors. The moment of conquest becomes the point in time in which conflicting experiences, and subsequently competing narratives of culture, emerge. However, when the memory of the defeat is forgotten and no longer painful, it may be integrated into the heritage of a people, as the legacy of the Roman Empire is now remembered in different parts of Europe.

These examples show how the past, and what constitutes a cultural inheritance, is a contested territory. These are the cases in which something that counts as a heritage for one group is simultaneously a symbol of oppression by another group. Every object in the city is the potential origin of a distinctive story, but simultaneously an integral part of a longer process that had started long before. It is a carrier of dormant stories that, if told from a different perspective, can reveal the power imbalances that were involved in its making and can be a potential source of tension and struggle. Based on the narratives that are constructed, the point in time that is used to mark the origin of a cultural item, or more broadly the cultural identity, is moved or ignored. All histories include these decisive moments, many of which have lost their poignance and divisiveness in collective memory, to form a multilayered palimpsest of different origins with diverse contributions to the resulting tapestry. Authenticity as a temporal reference is always a matter of choice among the overlapping shapes of this tapestry.

The conflicts over narratives and memories are not limited to the past. The present moment or the future events are also used as the temporal points of reference to claim authenticity. This is found particularly in the works of creativity and innovation, which, as we saw in the discussion of earlier chapters, are disputed. Can the design of a new place be selected as the origin for future authenticity? Can we call a completely new town an authentic town?

Urbanism is not a spontaneous creation from nothing, but it is an unfolding process, including a series of spontaneous events, intended actions, interactions, and intended and unintended outcomes. The present point in time is never detached from the previous moments and activities. The contested memories of the past become the constituent elements in the construction of the present and the future. As we saw earlier in this chapter, in the discussion of interrupted identities, the utopian ideas of a completely different reality are never free from the hidden and suppressed layers of the palimpsest.

As everything changes, the notion and expression of authenticity are also subject to change. Temporality means that things are always changing and so a fixed form of authenticity can never be established. The point of origin on which a claim to authenticity stands might be the description of a moment in time and not the next. Treating the urban heritage as an open-air museum tends to fix the past into specific moments with preserved characters, but the city is alive and continues to change and evolve. The stories told about the city are never-ending, and any attempt at freezing the city into a specific form may come into sharp contrast with the temporality of change.

## Authenticity and temporal origins

The temporal aspect of authenticity links a phenomenon with an origin in the past and develops an account that can reliably connect the present with the past and future. This exercise provides a sense of stability and continuity in a world that is changing at ever higher speeds and scales, with its associated sense of vulnerability and being dislodged and out of control. However, the selection of that point and the narrative that is produced for bridging the present to the past are matters of selection and interpretation, undertaken by some groups, using specific methods, and in specific contexts. Authenticity becomes both a state of being and a way of linking to the past, simultaneously with ontological and epistemological characteristics.

The discussion of the World Heritage shows how the concept of authenticity plays a significant role in establishing this line of connectivity to the past, in a context in which these connections are exposed to disruption and disconnection in periods of intensified urban development. It also shows how the account that constructs that connection may be produced by reliance on scientific or cultural sources of information. The discussion about identity and history also shows how these sources of information are always interpreted in a dialogue between the author and the remains of the past, between the present and the past. It is a process that starts with a question, embedded in a worldview, selective in its values, targets, and tools. The result may be satisfying for some, but disturbing for others, reflecting the inequalities in power between different groups. Authenticity in this sense, therefore, is the assertion of a temporal link to a point of reference in the past, in the present, or in the future, paving the way for a course of action; it is nevertheless partial and selective, and therefore, it should always be subject to critical examination.

# 7 Place

This chapter examines place as an origin, and authenticity as a relationship to that place. In place-related authenticity, something may be called authentic if it can be traced back to a specific place; its authenticity is only confirmed when that relationship is proven. This form of authenticity is based on assuming the integrity and coherence of the selected place, that is, its suitability to serve as a point of reference, an origin. Therefore, both the originality of the place of reference, and the reliability of the relation to that place are at stake. In other words, the place-related notion of authenticity requires demonstrating the authenticity of a place itself, as well as showing the authenticity of something through a proven relationship to that specific place. It is about the spatiality of authenticity, both in linking one place to another, which can be in another location or inside the same place, indicating a relationship between the inside and the outside. This chapter starts by examining the notion of authenticity as looking for an origin in space, for specificity and particularity, with authenticity as a desire to return to this specific place, such as the home or nature, where some originality and integrity is sought. This chapter also examines the practices of placing and distancing, which shape the dynamics of place in the diversified and stratified urban environment.

## In pursuit of an anchor in space

Authenticity is inherently a spatial relationship, as it qualifies the relationship between two places, either as distanced from one another, or as one inside the other, thereby the relationship between an interior and exterior. While time-related authenticity looks for a point of origin in the past or the present, place-related authenticity searches for a particular place as its point of reference. As everything has a spatial dimension, even the items in our minds and digital streams in electronic devices, in determining whether something is authentic, its relation to a place is evaluated. That place may be here, thereby making authenticity a claim to spatial continuity. The place of origin may also be somewhere else, hence authenticity becomes a relationship between two places, between here and there. The question is: how do we find that place? How can a place become a point of origin? What are the processes

DOI: 10.4324/9781003451778-7

through which this selection is made? Is this selection of a place as the point of origin a personal choice or a collective decision?

A starting question would be about the meaning of place, a question that first takes us to the differences between space and place, a key distinction in spatial arts and sciences. This distinction became the centre of attention after the modernist theories of architecture and urbanism made substantial changes to the urban environment (Sert, 1944). These theories and their implementation were criticized for not paying any attention to the specific features of the places and the lives of communities that were undergoing change (Gans, 1968; Jacobs, 1961). The mid-twentieth-century urban transformations were rapid and large-scale, driven by functional necessity and market rationality, rather than the social and cultural conditions of the urban contexts (Lefebvre, 1991). Attention to place was a critical reaction to the concepts of space that modernist technological functionalism advocated. However, the distinction between space and place has a much longer history.

In much of the world, and throughout history, the notion of place has referred to specific, relative, and limited spaces that bodies occupy in the world. In ancient Mesopotamian and Greek cosmologies, the earth was envisaged to be at the centre of a universe that was contained within celestial spheres with fixed stars and planets (Cornford, 1997; Feke, 2018). Even in this cosmology, space was finite, contained within concentric glass domes. A major point of departure, albeit limited to mathematics, was by Euclid, who conceptualized space as an absolute and infinite void, without a centre or circumference. For him, space was imagined as 'an immeasurable blank field, on which the mind could describe all the perfect figures of geometry, but which had no inherent shape of its own' (Cornford, 1976:5).

This abstract concept of space was revived after the Renaissance, when the earth-centred theory of cosmos with its fixed celestial spheres was shattered. A new paradigm emerged during the scientific revolution of the seventeenth century, which introduced a mathematical-mechanical image of the world. As a mathematician-philosopher, Descartes conceptualized space as an endless mathematical grid: 'a continuous body, or a space indefinitely extended in length, breadth, height or depth, which was divisible into various parts, and which might have various figures and sizes, and might be moved or transposed in all sorts of ways' (Descartes, 1973:103, *Discourse on Method*, 4). Newton, whose theories of physics became the cornerstone of modern science, envisaged space to be an absolute, independent entity: 'Absolute space, in its own nature, without regard to anything external, remains always similar and immovable' (Newton, 1845:77). For him space is absolute, true, and mathematical, which is free from any association with context and circumstance. In contrast, he acknowledges a subset of this space to be a relative, apparent, and common notion of space: 'Relative space is some moveable dimension or measure of the absolute spaces; which our senses determine by its position to bodies' (Newton, 1845:77).

Leibniz, also a mathematician-philosopher, objected to the absolute concept of space, arguing that space is the order of coexistences, thereby advocating a relational theory of space and time: 'I hold space to be something merely relative, as time is; that I hold it to be an order of coexistences, as time is an order of successions' (Leibniz, 1979:89). He explains this formulation:

> For space denotes, in terms of possibility, an order of things which exist at the same time, considered as existing together, without enquiring into their manner of existing. And when many things are seen together, one perceives that order of things among themselves.
>
> (Leibniz, 1979:89)

Despite this objection and the famous debate that went on, it was Newton's concept that prevailed for centuries.

The idea of place is a critical response to the mathematical idea of the universe that emerged in these conceptualizations. The abstract notions of time and space do not offer any specificity or distinction, in which something can be judged for its meanings and qualities. The world and its features are expressed in quantities, in the quantitative language of mathematics, which since Galileo has become the language of science. Modernist urbanism, with its admiration for science and technology, embraced this mechanical model of the world and the mathematical instruments of understanding and transforming it. However, when the application of these principles of modernist urbanism led to a destructive impact on the social fabric of cities without delivering its stated social ambitions, the idea of abstract space became a byword for domination and insensitivity (Lefebvre, 1991).

If space is infinite and described in mathematics, place refers to a part of space that is occupied by someone or something, finding meaning and value through its relationships with the surrounding environment (Goodall, 1987; Mayhew & Penny, 1992). Place is defined by 'felt value', security and stability, which is at odds with the openness and freedom of space (Tuan, 1977:3). Place is an embodied and embedded space, an expression of particularity, as distinctive from the generality and universality of space (Madanipour, 2003). As such, place has become a key concept in the emergence of a humanistic critique of modernist functionalism, which concentrated on delivering large quantities of utilitarian space. Functionalist reorganization of cities had been done in the name of addressing the acute urban housing problems, but at the cost of ignoring the existing social and cultural characteristics of human environments (Gans, 1968; Jacobs, 1961).

The conceptualizations of space run parallel to the conceptualizations of time. As we saw in the discussion of time, time and space became increasingly intertwined, to the extent that the idea of spacetime has been common in the physical and social sciences. It is not conceivable, as both Newton and Leibniz did, to envisage a space that is timeless, or a time that can exist independent of a spatial reality. Bergson (2007) complained that the notion of time has been understood through quantified, spatial concepts, while it should be independently understood as a qualitative phenomenon. But even qualitative time as experienced intuitively can never become fully free from spatiality that comes with the embodied nature of the human being.

In the notion of place-related authenticity, this integrated relationship between space and time is also visible. Authenticity draws on space and time, on geography and history. In everyday conversation, when we say something is authentically French, we mean that it has been made in, or has come from, France. It has not been made in another country and presented as French. Therefore, there is a geographical reference to a particular territory. This is what the winemakers do, marking on their bottles the country, region, and even the farm where the grapes were cultivated and wines produced. The certificate of authenticity here means a clear and factual correspondence to a particular territory.

This meaning of authenticity does not interrogate how something may be associated with that territory. Here the dimension of time and history come into play. The longer something is associated with that territory, the more authentic it is assumed to be. So when we say something is from Northumberland, it indicates that its origin is in that county, and furthermore seems to imply that this link is not fleeting and temporary, but it has been connected to that place for a long period of time. In other words, the certificate of authenticity becomes stronger with the passage of time. The longer something has been associated with a territory, and the older a territory has been established, the more authentic it is assumed to be. Space and time, therefore, are key ingredients of authenticity, with their nuances, which create gradients in the authenticity of something. This may give the impression of stability and originality, but it does not mean fixity and eternity. It indicates being anchored in a territory, with a slower pace of change, in a world that is rapidly changing.

The idea of place-related authenticity finds its anchor in place, as a locus of specificity, which can offer particular meanings to those who use it as a point of reference. The search for a point of origin cannot find a specific and meaningful node in space as an abstract void. This search needs to be able to pinpoint a location with a range of characteristics that can be traded in the marketplace of identities. Authenticity in relation to place is not necessarily referring to uniqueness, as it was the case in the works of art. There can be many authentic relations to a single place: for example, there is no limit to the number of authentic French restaurants. No single restaurant can claim to be the only authentic one to the exclusion of all others. The identity of a person, a group, a practice, a habit, or an object cannot be adequately described with the language of science. Science describes them in important and useful ways, but social recognition often depends on culturally meaningful descriptions, of which authenticity is an important one. This association with a place provides specificity and positivity, distinguishing something from others through what is thought to be an advantage. At the same time, it generates a sense of exclusivity, which can be a disadvantage to those who are thought to be outsiders. It produces a dialectic of inside-outside, inclusion and exclusion, with significant consequences.

## Particularity, difference, and similarity

The idea of place refers to particularity in space and time, as well as particularity in meaning (Madanipour, 2023). In phenomenological and humanistic critique of the modernist attitude, the place would offer personal and collective meaning (Tuan, 1977). Through long association with a particular location and community, persons and communities would feel bound to the place, interpreting and constructing meaningful relations with each other and with their environment. The place becomes an integral part of that community and the person's life story (Tönnies, 1957). It becomes associated with familiarity and continuity, a refuge from the restless space of flows (Castells, 1997), and a platform for rooted meaning among the forces of alienation and rootlessness (Heidegger, 2011; Norberg-Schultz, 1980). This desire for rootedness is a direct response to the urban industrial processes that uprooted people from their towns and villages, bringing them together in large and alien cities (Briggs, 1968), and displaced them once again when these industries declined through neoliberal globalization. As the pace of life has accelerated, mobility has increased, and individualization has intensified, the desire for safety and security has been felt more than ever before. However, the emphasis on the place may not suffice for overcoming these forces of fragmentation and diversification, remaining only an ideal that could never be fully realized.

At the same time, this emphasis on rootedness can lead to a process of excluding those who are not thought to be a part of the place, triggering a discourse of othering, which can lead to prejudice, exclusion, and even violence (Madanipour, 2020). It is not difficult to see how the extension of Heidegger's notion of authenticity can lead to advocating 'participation in an exclusive community' and show 'a worrisome proximity... to Nazi ideology' (Zuidervaart, 2007:87). This is what many nationalist and populist discourses have advocated, asserting their sense of place through demanding the forceful exclusion of immigrants and outsiders. Nations and their territories have been associated with much conflict and bloodshed while remaining ambiguous in their meaning (Hobsbawm, 1992). It has been argued that 'It is nationalism which engenders nations, not the other way round' (Gellner, 1983:55). These 'imagined communities' have released the forces of intolerance and reactionary politics, but they have also triggered emancipatory, anti-colonial movements (Anderson, 2006).

As Edward Said (1988) argued, this could lead to new claims for authenticity, in the form of religious fundamentalism, but also in more critical and emancipatory forms, in 'the interests of the unheard, the unrepresented, the unconnected people of our world, and who do so not in "the jargon of authenticity" but with the accents of personal restraint, historical scepticism and committed intellect' (Said, 1988:60). In using the title of Adorno's book on the critique of Heidegger, Said distinguishes two forms of authenticity, reactionary and emancipatory, acknowledging one and keeping a distance from the other. The search for authenticity has been a rallying cry for many post-colonial thinkers but also a cause for closure, rigidity, and hostility to new ideas and practices (Lee, 1997).

Therefore, nations and nationalism, as one of the powerful discourses that link people to places, have had a dual character: constructed from above, but also from below, drawing on the needs and aspirations of the people, which could not even be related to nations and nationalism (Hobsbawm, 1992:10). Within the last 100 years, at least two major movements have imagined that the question of nationalism has been solved: the Soviet internationalism, which claimed to have solved the problem of ethnicity and nationalism through solidarity and collectivism; and global neoliberalism, which embraced the belief that with intensified mobility the age of cosmopolitanism has arrived. However, as both movements came to face their limitations, what reemerged was a resurgence of nationalism and a discourse of exclusion against migrants and minorities.

A place can offer a strong, well-recognized, and stable descriptor for self-identification. It can also be a pathway to stigmatization and humiliation. If your place of birth, residence, or work is recognized by others as ordinary or highly regarded, you will be proud to announce your origins and affiliations with that place. This is why national and regional identities are often used as one of the indicators of authenticity. However, if your reference place has a negative image in the mind of others, such as particular parts of countries or particular parts of cities that are associated with poverty or crime, place-related identity becomes stigmatizing. For many residents of poor neighbourhoods, the name of their street or neighbourhood can become a stigmatizing obstacle in their social interactions, blocking the opportunity of receiving respect from others or to be offered a job. Place-related authenticity would mean describing a person or a group as having originated from that area, but the negative image of the area generates gaps, tensions, and alienations.

If your authenticity is defined through your rootedness in a particular place, and if others consider that place to be somehow socially inferior, you may go through a process of alienation from the self and from the place, with all the negative social and psychological consequences that may result. Looking for a place of origin, therefore, is not necessarily a path to peaceful reunification with the self but could lead to a painful experience of being singled out and humiliated. Rather than embracing place-related authenticity, what follows may be engaging in play-acting, creative narratives, inventing stories, aggression, evading tactics, and passive introversion, just for overcoming the pain of stigmatization. It creates a dialectic of inside and outside, concealment and disclosure, inclusion and exclusion, with far-reaching implications for society and space.

Place signifies particularity and specificity, a key part of the narratives of identity and authenticity. If, as Ricoeur (1991) argues, narrative identity is about the stories that we tell about ourselves, the place and our relationship to the place are integral parts of that narrative. This narrative looks for particularity in space and time, and asserts singularity and distinction in meaning. However, identity is always a relation of similarity and difference, generating a diversity of narratives told by a multiplicity of voices (Jenkins, 1996; Madanipour, 2013).

When place-related authenticity refers to continuity within the same place, it emphasizes continuity through time, downplaying or rejecting the ways in which the place has changed. A real or imagined place serves as its own origin, as its inner self that has now been masked and would need to be recovered and revived. This can lead to different recipes that range from the recovery and protection of cultural heritage to the reactionary politics of rejection and exclusion, from searching for a dignified identity to mobilizing aggression towards others.

When place-based authenticity refers to somewhere else from which something is originated, it is a horizontal, equal relation of simultaneity between two places, but a relation in which one serves as the origin for the other, thereby privileged by higher significance and by temporal priority. Place 'a' should exist before place 'b', so that 'a' can be considered an origin for 'b'. Their relations are also hierarchical: place b becomes distinguishable by its relation to place a, thereby a is given priority in significance over b. Therefore, by serving as its origin, place a is temporally prior to, and more significant than, place b. Meanwhile, place b's identity carries within it a trace of place a.

Reference to place as a point of origin emphasizes its uniqueness, and how it is different from other places. Through reference to the uniqueness of a place, a stronger case seems to be made about the identity of something. As the world becomes more urbanized and globalized, it loses many of the local distinctions that made up its cultural diversity and richness. Many commentators have complained about the lack of identity in large modern cities, where all buildings and roads look the same, all following the same planning ideas and technological instruments, where modes of living are increasingly similar, where similar products and services are present, and where city-dwellers seem to be no more than cogs in a giant machine. In response, efforts for generating and claiming distinction are made at all levels. The drive for authenticity appears to be a critique of these similarities, looking for originality in distinctive and unique places. A point of origin is assumed to be a singular one, quite unlike any other, so that it can offer a sense of stability and confidence.

However, there is always a multiplicity of narratives and narrators, which can potentially turn every story and every place into an authentic one. For example, Venice is a unique city, with its particular location, composition, and materiality. When other cities reproduce a version of Venice, such as Doha or Las Vegas, these copies are seen as fake, where only a representative image of the city is produced, without the authenticity of the original (Figure 7.1). In the original Venice, every day thousands of tourists visit the city to experience this uniqueness, walk along the canals and take a gondola trip around its narrow and winding water alleys. But as Italo Calvino's book *Invisible Cities* shows more than half a century ago, this unique city can find many faces and interpretations, even for a single Venetian like Calvino. In the book, Marco Polo describes for the Great Khan of China dozens of cities that he has seen, but as he says to the Khan, 'Every time I describe a city I am saying something about Venice' (Calvino, 1997:86). Not many people will have the poetic imagination and richness of Calvino's storytelling, but if this multiplicity of interpretations is multiplied by the number of people who live in and visit the city, the unique Venice becomes millions of Venices, each associated with a different experience and memory. While Venice remains physically the same, albeit slowly declining, its experiences and interpretations linger and multiply, as shown by Calvino's book. When taking into account the famous Venetian masks that portray the city's population in a thousand different ways, we are faced with potentially endless versions of the city's representations of people and places (Figure 7.2).

*Figure 7.1* The stereotyped image of a city can have many copies around the world (Doha, Qatar)

*Figure 7.2* The original city, meanwhile, has many faces and countless interpretations (Venice, Italy)

However, identity is not merely an indicator of difference but also a reference to similarity (Jenkins, 1996; Madanipour, 2013). On the surface, the global urban competition looks for uniqueness through iconic architecture. But in doing so, this competition is an attempt to become the member of an elite club of cities that are rich and important enough to construct such iconic buildings. Cities are ranked for their distinctive characters, which is why many city authorities promote the construction of iconic works of architecture so that they can stand out in the global competition for investment and tourism. Famous architects and unusually shaped buildings are the symbolic instruments for claiming that distinction. But underneath, there is a desire for inclusion in a group that shares a similar approach to symbols. In addition to the similarity in approach, these cities are subject to the same global forces of capitalism, which turn their specificities into commodities, thereby reducing their distinctiveness. The presence of the same global brands in many cities has led to a campaign against 'cloned cities'. The drive for distinction runs against the background of the strong economic and technological forces of assimilation.

Authenticity may imply continuity in time but distinction in space. In time, it means how something keeps the lines of continuity with a past origin, hence maintaining a degree of cultural heritage that would show how the present bears some traces of the past. A narrative of similarity is created in which the present culture is somewhat similar to a selected point of origin, whether an event, an object, or a culture. However, our interpretation of the past is an invention in the present time, partly a reflection of our current subjectivity, rather than purely based on how something existed in the past, a past that is not accessible to us. We need to acknowledge the limits to the information that is available about that origin and what it meant at the time. We also need to acknowledge the changes that a place has gone through, as well as the changes in interpretation through ages. Therefore, authenticity in time would also include the differences between now and that selected point of origin in the past. Furthermore, a sense of authenticity might lie in a break with the past, as the sign of something new, which locates the spatial and temporal origin to here and now, referring to both continuity and change.

In space, descriptions of identity show how a place is defined by its differences from other places, that is, how it is a distinctive space. In this account, some features of a place are selected to show its uniqueness, its originality and authenticity. However, these distinctions are not fixed and stable, and they may change over time. Moreover, searching for distinction in space would also show the similarities between a place and some others. In other words, the comparison between two places will always show their similarities and differences. Place a and place b are never exactly the same, and never completely different. Their patterns of similarity and difference generate a series of features that are used to characterize each place. Emphasis on distinction would ignore the many ways in which two places may be similar. Therefore, authenticity in space would indicate both similarity and difference.

Temporal authenticity may be considered as continuity and similarity, and spatial authenticity as difference and distinctiveness. However, as similarity and difference are always entangled, as space and time cannot be separated from each other, and we are always thinking in spacetime, rather than space and time in different ways, this interpretation of authenticity would need to be enlarged. Authenticity as the interpretation of a spacetime event will always combine similarity and difference, a relationship between that spacetime event and other spacetime events.

## Returning home

The idea of the self, as the platform on which authenticity is often built, is not detached from the world but is rooted in specific places. Home is one of the most important of these specific places, where the notions of roots, origin, and identity are intertwined.

Heidegger distinguishes between housing and dwelling: 'The truck driver is at home on the highway, but he does not have his lodgings there' (Heidegger, 2011:244). Through linguistic analysis, Heidegger shows the deep links between building and dwelling: 'For building is not merely a means and a way toward dwelling – to build is in itself already to dwell'; it 'signifies to remain, to stay in a place' (p.244). Building and dwelling are also related to being: 'To be a human being means to be on the earth as a mortal. It means to dwell' (p.245). Building in a narrower sense refers both to cultivating and constructing and in the habitual ways in which these activities take place, the proper meaning of building as dwelling, 'as being on the earth' is forgotten. The fundamental character of this being on the earth is to be free, at peace, and protected. Therefore, 'To dwell, to be set at peace, means to remain at peace within the free, the preserve, the free sphere that safeguards each thing in its essence' (p.246). It spares and safeguards them: 'the basic character of dwelling is safeguarding' (p.247).

For Heidegger (2011), buildings 'gather' and unite the elements of the world, which he poetically calls the fourfold of earth, sky, mortals, and divinities. This is the role of 'things' as the mediums of bringing the different dimensions of the world into a unity, creating a 'locale', which is similar to what the humanistic literature has called place, and prefigures Arendt's (1998) discussion of the public space. It is different from space in the abstract sense of distance and extension and sets the condition of staying, or being on earth: 'building, by virtue of constructing locales, is a founding and joining of spaces' (Heidegger, 2011:252). Building, dwelling, being, and thinking are all connected. This is why Heidegger argues that the 'plight of dwelling does not lie merely in the shortage of houses', but it 'lies in this, that mortals ever search anew for the essence of dwelling, that they must ever learn to dwell' (p.254). With his example of a 200-year-old farmhouse in the Black Forest, Heidegger delivers his anti-modernist message about what it means to dwell and how it has been forgotten in the modern world, causing humans to become homeless. For him, human homelessness lies in not thinking of 'the plight of dwelling as the plight'; therefore, 'as soon as man gives thought to his homelessness, it is a misery no longer', as it would call humans 'into their dwelling' (pp.254–55).

As the champion of authenticity and a harsh critic of modern technology, Heidegger's conservative message of rootedness in place is his response to the alienating and fragmenting forces of capitalist industrial society. This is a message that had run, in various forms and guises, by the revolutionary and romantic movements against the ills of the new society. Rousseau, an early advocate of authenticity whose ideas inspired the French Revolution, contrasted the urban society of his time to the idyllic and integrative rural life, in which the division of labour and fragmentation of life into different roles and functions did not exist. Rousseau describes the life of a Swiss village, where farmers had all the skills necessary for maintaining and repairing their own houses, where they lived a happy life of singing, dancing, and playing the flute, which he thought was a model of free and egalitarian life (Cranston, 1968:18). The disdain for the fragmentation and sterility of the modern urban life is also evident, albeit in a completely different direction, in the work of another champion of authenticity, Friedrich Nietzsche.

Nietzsche's fictitious Zarathustra, the Persian prophet, does not have a home to return to: 'But a homeland I found nowhere: unsettled am I in all settlements and a departure at all gates' (Nietzsche, 2005:105, II, 14). His shadow complains, ' "Where is – *my* home?" That is what I ask and search and have searched for, and have not found' (Nietzsche, 2005:239, VI, 9). He is tired of life, disgusted by its repetitiveness: 'And eternal recurrence even of the smallest – That was my loathing of all existence!' (p.192, III,13). In this life, 'all things return eternally'; everyone would return '*not* to a new life or a better life or a similar life', but 'to this self-same life' (pp.192–193). As a demon once put to a lonely Nietzsche, what would you do if you are told that, 'This life as you now live it and have lived it you will have to live once again and innumerable times again; and there will be nothing new in it' (Nietzsche, 2001:194, III, 341). In this case, 'The eternal hourglass of existence is turned over again and again, and you with it, speck of dust!' (Nietzsche, 2001:194, III, 341). When faced with this question, would you despair or would you find this insight 'divine'?

The Nietzschean return is emphatically expressed as a return to the same life; but this is not a literal reincarnation. Returning to life in Indian traditions of reincarnation is a return to life but in a different state, on a trajectory that could descend or ascend, driven by the ultimate promise of no return (Rahula, 1974). The Nietzschean return is not like the film Groundhog Day, where an exact replica of this life is repeated; nevertheless, it is a reference to a mental state of finding daily life repetitive, reliving the sort of experiences that you would have in this life over and over again. This is shown in Nietzsche's self-reflection: 'You will never pray again, never adore again, never again rest in endless trust; you refuse to let yourself stop to unharness your thoughts before any ultimate wisdom, goodness, or power....' (Nietzsche, 2001:161, No.285). In other words, you may wish to renounce many things you feel and do in your life. But he warns: 'Man of renunciation, all of this you wish to renounce? Who will give you the strength to do so? No one yet has had the strength!' What is needed is the strength to live confidently: 'perhaps man will rise ever higher when he no longer *flows off* into a god' (Nietzsche, 2001:162).

Rather than a place, a time, or an authority to return to, Nietzsche's Zarathustra finds a home in his own life, finding strength and meaning in the endless circle of events. He becomes a teacher of the eternal return. He is encouraged to 'go out to where the world awaits you like a garden', with its 'roses and the bees and the flocks of doves' and 'especially... the songbirds' (Nietzsche, 2005:192). With this celebration of life, he is warned that he is and must become '*the teacher of eternal recurrence*' (Nietzsche, 2005:192). This is simultaneously a return to the everyday and rising above it.

If in Freudian psychoanalysis childhood is the temporal origin of the psyche, the home would be its spatial origin, where childhood unfolds, and where large parts of what becomes the self are formed. The idea of authenticity as returning to one's true self inherently involves a return to home, where the person originated from and where life was probably at its most integrated form. Returning home would then mean regaining the authentic self. The story of return is an integral part of the experience of movement, in space or time, away from home or from an earlier stage of life. What if the home was never a warm place, but always a place of tension and violence, or one which has changed so much that it is no longer possible to recognize it? Assuming permanence in the home is as misleading as assuming fixity in the self. The combination of the never-ending desire for return and the impossibility of returning to the past is so fundamental that it is told tirelessly in various forms throughout centuries.

The ancient Greek hero Odysseus, the king of Ithaca, wins the long war with Troy and sets to return home (Figure 7.3). However, as Homer's epic mythological poem *Odyssey* describes the journey, it takes ten years of wandering in the seas for him to get back to his point of origin. Homer's poem opens with describing Odysseus a 'resourceful man, who wandered far and wide'; this was a struggle for return but also included discovery and learning: 'many men's townships he saw, and learned their ways of thinking, many the griefs he suffered at heart on the open sea, battling for his own life and his comrades' homecoming' (Homer, 2018:27). The obstacles on the way are varied and enormous to overcome, but he survives and eventually returns to reclaim his kingdom and to unite with his wife and son, who are under pressure from her suitors. Homer's poem, written in 24 books in the eighth century BCE, is considered one of the greatest works of literature. It is a story of separation, desire for return, temptations, and impediments posed by the natural and supernatural forces on the way. It also highlights courage and resourcefulness in fighting many battles for overcoming these obstacles, and finally the fortune of reclaiming a rightful place at home.

The supernatural forces and the misadventures that prevent Odysseus from returning to Ithaca cannot persuade him to give up the attempt. Home may be thought of as a safe place for a return, but for Odysseus, home is as dangerous as any of the towns that he has visited in his perilous journey. But even the dangers at home are preferable to the possibility of safety elsewhere. In this ancient culture, travel across the seas and establishing new settlements are an integral part of life, but the significance of the home in a travelling culture becomes even more crucial. Attachment to one's home is thought to be worth all the sacrifices, as exemplified by Socrates who preferred death in Athens to an exiled life in other cities (Plato, 1997, *Crito*, 52c). In this context, home is a fixed point of reference, a basis of identity, and a powerful magnet pulling everyone towards itself, urging them to return.

The story of return also forms the opening lines of *Masnavi*, the masterpiece of the thirteenth-century Persian poet Rumi, the mystic founder of the whirling dervishes who is buried in Konya in modern-day Turkey. His famous song of the reed is about a reed-flute that is cut off from the field, and since then it has played a sad song of loss, and a heartfelt desire for return. Rumi (1983:1) writes: 'Listen to the reed-flute (*Ney*) as it tells a story, complaining

*Figure 7.3* On his way back home, Odysseus, in this statue from first century BCE, looks back to see if someone is following him (Athens, Greece)

about separations… Whoever remained far away from their origin, will seek a time of reuniting with it'. This story of return is not to a worldly home, but to an origin in the spiritual realm, where the human soul has come from and to which it desires to return.

The gap between the two worlds and the desire for return is also the basis of the biblical story of Adam and Eve who were expelled from Eden, resulting in an eternal yearning for return. The entire course of life becomes a struggle for return, which may only be possible at the end of the world. The story of return refers to a gap between two worlds, between the materiality of this world and the innocence of the origin, and the journey of life that is the struggle of passing through these worlds. The story of return is also told as a promise of justice and hope for emancipation: the return of a saviour to the world to bring justice, and the return of everyone from death to face justice on judgement day. With the decline of this hope, the only hope that could remain would be a return to the self in this life.

As the person and the home both evolve and change, and as the world around them is transformed, a return to home is never a return to that origin, even if this return is all that is desired. The separation from home and the desire for return express spatial and temporal gaps that a person wishes to bridge and mental ruptures to mend. If the unfolding temporality of childhood combines a growing independence with psychological gaps that may never be filled, the unfolding spatiality of the home expands the growing child's horizons but inserts distances and ruptures in memories and relations. These gaps and ruptures tend to have been caused by forces that are beyond the person's control: the powers that displace persons from their home, the obstacles that are on the way back, or the person's internal conflicts and circumstances. The home becomes the spatial expression of a past, an inaccessible origin to which a return is never fully possible.

Meanwhile, the desire for return may persist, but not be supported by the will to return. In what has been called the 'myth of return', many immigrants are thought to maintain a longing for a return to their birthplace, but for a variety of educational and economic reasons never act on it (Anwar, 1979). This would apply to those who have migrated from towns and villages to large cities in the same country or those who have crossed the borders to other countries and continents. Whether by force or by choice, those who have left keep the memories of their childhood and youth as precious parts of their identity; they yearn for a return to those memories and to a recovery of the self, which are no more than memories, as the conditions that make them possible are no longer present. Having lost their old home and not feeling completely at home in their new places, they seem to remain suspended in between two worlds, where authenticity as a return to a place becomes impossible.

These brief stories from psychology, literature, religion, and philosophy show displacement, self-alienation through separation from an origin, yearning for return, and dreaming for reunification with the origin. The journey of life and its adventures and circumstances have caused this separation and alienation, but the promise of authenticity seems to lie at the end of this road, to be achieved through reunification with the origin. The inherent assumption in the stories of return is the stability of the home as well as the person who wishes to return. However, the person, the home, and the world that framed that original state have all changed, and a return, if it happens, would be by a different person to a different place. While the desire for return may remain strong, the authenticity of an imagined reunification may remain beyond reach and the distance between the two worlds may be difficult to bridge. At the same time, these are not simple stories of sadness and loss, as it is possible to interpret them as having gained two worlds. The burden of bridging the gap between these worlds is always heavy, but it offers possibilities for constructing new forms of authenticity that are relational and future-oriented.

## A place in nature

The search for authenticity has frequently led to the natural world, where simpler, more original, and more integrated forms of life are sought. In contrast to the complex layers of convention and compromise that constitute the human world, the natural world is thought to be purer, nearer to the essence of the world. All that is needed, it is thought, is removing these human-made outer layers to get to the truth, finding everything in its authentic form. The natural world within the human body and the natural world outside it, in its various forms on earth and beyond, have both been points of reference in this search for authenticity (Figure 7.4).

The word nature has been the subject of much debate. As Raymond Williams (1985:219) suggests, 'Nature is probably the most complex word in the language'. He proposes three meanings for the word: '(i) the essential quality and character of something; (ii) the inherent force which directs either the world or human beings or both; (iii) the material world itself, taken as including or not including human beings' (Williams, 1985).

*Figure 7.4* Within the human body and in the outside world, nature is thought to be the original source of authenticity (Duns, UK)

The first meaning underlies the scientific endeavours that throughout history have attempted to uncover those essential qualities, as embodied in the laws of nature. The second meaning, which assumes an inherent force, has long been the basis of supernatural beliefs, which assign power and agency to the material world. The third definition, especially when the natural world is defined in contradistinction to the human world, has fuelled the claims of human superiority over the rest of the world. The long-term use, the diversity, and even opposite meanings of the word have been at the root of its complexity and ambiguity.

The Oxford English Dictionary provides two distinctive meanings that are in line with Williams' first and third definitions. In the first sense, it refers to 'The phenomena of the physical world collectively, including plants, animals, the landscape, and other features and products of the earth, as opposed to humans or human creations'. However, humans are also a part of this physical world, blurring the lines of distinction. This is why the expression non-human is inherently an anthropocentric expression, while apparently an attempt to distance itself from anthropocentrism.

In the second sense, nature indicates, 'The basic or inherent features, character, or qualities of something'. In some discourses, this indicates the essence of things, fixed and unchanging, with universal and eternal laws that need to be discovered. At the same time, natural events have long been seen as unruly and unpredictable, which shows a paradoxical approach to the natural world, seeing it as both orderly and disordered. Meanwhile, anti-essentialist critics have questioned any such fixity, especially in relation to human society, and some have even questioned the use of the word nature as appropriate for describing anything. The two meanings, physical world and basic features, expand the word horizontally and vertically, covering the world as a whole and what lies beneath its surface. They have been used interchangeably and simultaneously in both meanings, referring to both surface and depth, both the multiplicity of ever-changing phenomena and the fixity of some inherent laws and regularities.

The word nature, which is a key reference point for authenticity, is therefore a field of ambiguity, displaying paradoxes that cannot be easily addressed. Nature is considered to be an ordered system, with universal laws that may be hidden but can be discovered by science, but at the same time, nature is viewed as wild and unpredictable, at the intersection of powerful forces that cannot be grasped or tamed. Humans are considered to be outside nature, ruling over it and exploiting its resources, but at the same time, humans are parts of this natural world, subject to its processes of growth and decay. The place of humans both inside and outside the natural world, the lawfulness and unruliness of the natural world, and the presence of some essence or the absence of any essence in nature are all paradoxes and dichotomies, which turn any roadmap towards authenticity too complicated as a travel guide. Meanwhile, there have been paradigmatic changes in the approaches to the natural world, from a place of harmony and order to a rough wilderness, a bottomless source for exploitation, and finally a vulnerable environment that needs protection.

For Renaissance artists, nature was a place of order and harmony; an appeal to nature meant finding a source of inspiration for beauty and perfection. They looked for perfect proportions in the human body, which they drew through the use of geometry in their designs for buildings and cities (Alberti, 1988; Filarete, 1965). This was exemplified in Filarete's design for Sforzinda, the first ideal, star-shaped city of Renaissance, which was 'the first wholly symmetrical town plan in Western history' (Pevsner, 1963:185). His ideal proportions were based on the size of the human head, and its relations to the rest of the body. As 'the most noble and most beautiful member', the human head was the measure of the body: 'if the arms are opened and the hands extended, [the man] will be nine heads in either direction' (Filarete, 1965:8). In turn, the body was the measure of an anthropomorphic world. Using the human figure as its inspiration would ensure that his city would be 'fine, beautiful, and perpetual according to the laws of nature' (Filarete, 1965:5). Humans were a microcosm, and as God created the world in His image, the designer could now create a city in the godly image of the human (Madanipour, 2022b).

As Filarete (1965) explains in his *Treatise on Architecture*, which he wrote in 1461–1464, he carefully located his utopian city in an idyllic landscape, but he was also concerned about the potential impact that his city would have on the natural environment. This was a search for harmony and balance between the human and natural worlds. However, the selection of the human body as the point of reference for the natural order showed an explicit preference for the human world, which was strengthened by the discovery of the laws of perspective in around 1425, empowering the spectator's view of the world. This design took the shape of central composition, placing a single centre in the middle of a building or a city, as Filarete did, and as Alberti (1988) had advocated before him. As Pevsner (1963:182) writes, 'a central plan is not an other-worldly, but a this-worldly conception'; in centralized buildings, 'The building has its full effect only when it is looked at from the one focal point. There the spectator must stand and, by standing there, he becomes himself "the measure of all things"'.

The origin of the designs for the human world was thought to be the natural order, but locating that order in the human body elevated the human to become the centre of the world. The passage from the medieval theocentric worldview to the modern human-centred worldview would be expressed in this way in architecture and urbanism, as well as in the rise of absolute monarchies in Europe (Machiavelli, 1950; Madanipour, 2007). This was an early sign of the divide between these two worlds when the human world sets out to become the centre of the universe and conquer the natural world, a divide that would trigger the search for authenticity in future centuries. Meanwhile, the human body was an integral part of the natural world, generating a paradox for the future search for authenticity, which would eventually return to the human.

The mechanical sciences that emerged in the early modern period looked at the natural world as an ordered totality with fixed laws, which needed to be discovered and tamed (Hollis, 2003). These natural laws were to be found in both the physical and the social worlds. At the same time, they invented an imaginary 'state of nature', which was the opposite of their own society, with a radical otherness that would justify what they envisaged as their civilizing efforts (Thomas, 2020). For the early modern thinkers such as Hobbes, the state of nature was a place of savagery and conflict, where a 'war of every man against every man' was raging and where life was 'solitary, poor, nasty, brutish, and short' (Hobbes, 1929:96, Chapter 13, 62–63). His idea of the state of nature is not about ancient human society but reflects his own contemporary society, which was going through a civil war, political legitimacy was lost, individualism was rising, and the nation-state was emerging. This is why for him nature is a place to avoid, rather than a desirable destination.

Similarly, for Descartes, nature was something to conquer, rather than celebrate, advocating to 'make ourselves, as it were, masters and possessors of nature' (Descartes, 1968:78). In his dualistic thinking, mind and body were completely separate, and it was necessary for the mind, which represented human rationality, to control the body, which was a part of nature. The conquest of nature was the conquest of the body by the mind, a shift from the integrated body of the Renaissance to a subjectivist mindset. This was also the time of the conquest of other lands around the world by the emerging European powers and the exploitation of natural resources at the service of profit and prosperity. For the early modern colonialists, as John Locke argued, property rights were natural rights, and nature only found value when human labour was at work. Without human intervention, land 'would scarcely be worth any thing', but it would find value when transformed by human labour (Locke, 1821, II, para 43).

The idea of nature, however, was radically changed through a critical reaction. The state of nature became an imagined state of purity and peace lived by early communities, an idea which served as a point of origin for the emerging romantic movement from the middle of the eighteenth century. As urbanization and industrialization grew, some thought that the pains of modern capitalism could be remedied by a return to those simple forms of life, which were imagined to be free from inequality and suppression (Marx and Engels, 1985; Rousseau, 1923). The mythology of origins and the desire for a return to natural and simpler ways of life were behind the romantic and revolutionary movements that swept across Europe and its colonies around the world. Meanwhile, the European encounters in voyages of discovery and conquest with communities that lived in harmony with their natural environment paradoxically suppressed or destroyed these desired forms of life and transformed their natural habitat.

The romantic movements since the eighteenth century have fought against what they saw as the dangers that industrial urbanization poses to the natural world. Their response was so strong that a dichotomy of reason and emotion has become a central theme of the culture, a dichotomy in which the notion of authenticity is located. If rationalism and functionalism have turned the natural world into a mere source of economically exchangeable assets, naturalism and the ethics of care look for the intrinsic values of the natural world (Madanipour, 2022a). Nature provided an origin on which authenticity as one of these intrinsic values could be based. The natural world was no longer considered to be dark and uncivilized, but pure and original, offering an escape from the city. The English romantic poets Wordsworth and Coleridge took refuge in the Lake District, and Henry David Thoreau wrote: 'Let me live where I will, on this side is the city, on that side the wilderness, and I am ever leaving the city more and more, and withdrawing into the wilderness' (Quoted in McKusick, 2000:2). The escape from the city was industrialized through suburbanization, which started in the nineteenth century with the invention of trains and accelerated by the motor car in the twentieth. The romantic escape from the city also inspired the modern environmental movement, as testified by the editor of a collection of papers on green studies: 'Sometimes we may have the feeling that we are only just catching up with Wordsworth' (Coupe, 2000:6).

For evolutionary psychologists, the main patterns of human behaviour were formed during the millennia of life as hunter-gatherers who lived in small mobile bands, which was the longest form of life for the *Homo sapiens*. Life in agrarian settlements for the past 12 millennia is a relatively short episode in the history of humanity, and life in high-density industrial cities is even a much shorter episode. In this periodization, the search for authenticity as an original temporality may look for origins in the hunter-gatherer life, where the deep roots of human beliefs and behaviour lie. Accordingly, anything afterwards, especially urban living, would be considered expressions of inauthenticity, something that is at odds with humanity's long-established physical and mental habits. Even though these researchers may not use the concept of authenticity, they often refer to the illnesses that are caused by a mismatch between our deep-rooted features as formed during the millennia of foraging, and our industrialized way of life in urban centres. While a return to nature may seem to require a recognition of these deep habits, it may not be able to change the circumstances that are needed to overcome this mismatch. The desired authenticity of uniting with a simpler form of life and a return to the innocence of a natural habitat arise from a form of urban life that has been detached from these conditions and may not be able to return to them.

With climate change and the emergence of the Anthropocene, the concept of the natural world has dramatically changed once again, at least for those who are aware of the risks facing the future of life on earth. The diversity and even the survival of various life forms are under threat, and the future of human settlements in many parts of the world is in doubt. Rather than a wild force that rules over human lives, or a bottomless source for exploitation, it is a vulnerable earth that needs protection and care. In these circumstances, most design and development efforts claim to pay attention to this need, and keywords, such as sustainability, resilience, and green development, have become an integral part of the professional lexicon. In this context, the emphasis on naturalism has arisen, looking for a revival of the endangered species and protecting the threatened habitats. At the same time, as climate change accelerates, the return to those conditions becomes increasingly more difficult. Authenticity as a return to nature may not be feasible in an urbanizing world, where most of the world's population now lives in urban areas. If suburbanization once was thought to be a compromise between the city and the countryside, by now it has become a byword for environmental degradation. The genius loci, which was thought to represent the spirit and essence of a place, has now been so deeply hidden beneath the layers of topographical transformation that a return to such originality would no longer be within easy reach. Rather than a natural landscape, it is almost always a cultural landscape that is the context of urban design and development. A return to nature is always mediated via culture and society.

## Placing and distancing

With the decline of modernism, the term place found widespread popularity, which remains to this day, as exemplified in the notions of neighbourhood, place-based strategies, localism, and placemaking, all reflecting a preference for particularity and distinction. However, place and particularity are not always positive attributes that are used innocently. In this section, I use two concepts of 'placing' and 'distancing' as a critical perspective on the ideas and practices of distinctive neighbourhood development (Madanipour, 2021).

An example is from a housing development in an area called the Great Park, located on the northern edge of the city of Newcastle upon Tyne. Its development started in 2001, delayed by the economic crisis of 2008, and when it becomes eventually completed, it is planned to have 4500 residential units, split into several neighbourhoods. In many ways, it is a typical British suburban housing development, mainly formed of two-storey houses with gardens, built by mass housing developers, for middle-class households with a high level of car ownership; the type of development that has been going on for decades. A noticeable difference is the presence of some apartment buildings, which are mixed with suburban houses, at a higher density, and in a 'new urbanist' layout. This development has been promoted as a sustainable development, and a thriving community made of distinctive neighbourhoods. Its website claims that, 'Newcastle Great Park isn't a collection of buildings. It is a collection of homes, businesses and people who come together to live, work, learn, collaborate and smile'. But behind this promotional façade, this suburban development suffers from many of the problems that have been associated with such areas, such as lacking the basic social infrastructure, so for most needs people have to travel by car.

An action group made up of some local residents has called the Great Park the Great Betrayal. They complain that the claims made by the consortium of house builders who have managed the process of development have not been met, and so they warn the potential buyers to stay away. This shows a well-known gap between rhetoric and reality, between the claims that are made in the development process, and what has been delivered in practice. It shows a temporal and economic gap between the buildings and infrastructure, between what can be sold for profit, and what is a necessary but not financially rewarding cost. The first is speeded up, and the second is delayed.

In particular, two techniques of planning and design in urban development, and more broadly as techniques of social organization, that I want to emphasize here, are placing and distancing. The area is placed on the northern edge of the city, next to other middle-class areas. This generates a safe distance from the city, with all its social and environmental problems. Placing and distancing become the operating mechanisms of the land and property markets and of the formation and maintenance of a social geography. These two techniques are key tools for social organization, which are deployed in both public policy and market operations. Placing envisages social groups as distinctive categories, allocates them a place in the social hierarchy, and consolidates this hierarchy through spatial organization, where each group has a place in the social order and the urban landscape. Distancing keeps these groups apart, maintains their distinctions, and consolidates the social arrangement in the form of a fairly stable spatial pattern.

Around the same time that the Great Park started to be built, another major programme of urban development started in the city, which was branded Going for Growth. The city's mayor invited the late star architect Richard Rogers to redesign the city and plan for its future. According to this plan, the city was envisaged as a series of distinctive neighbourhoods. Thousands of houses were going to be demolished and many more to be built. This plan was shaking the order of placing and distancing that had been operating in the city for generations. But this old order was no longer functional, as the process of deindustrialization had changed the economic foundations of the city and hollowed out the working-class neighbourhoods. This was a plan of displacing and replacing. This process of placing and displacing, which reflects the structural economic changes in society, has not been new in the city. It had been at work during the large-scale, modernist redevelopments after the Second World War, when many old working-class areas were branded as slums and were redeveloped with serious social consequences.

At the time of the policy of Going for Growth 20 years ago, however, people in these Newcastle neighbourhoods could not accept such wholesale displacement. Protests and campaigns successfully defeated the plan, and Richard Rogers was dismissed by the city. In its place, they initiated several community participation programmes to engage people in the city's new, more modest plans. These two examples from Newcastle, a suburban scheme and an inner-city plan for renewal, show the complexity of urban development, and how there are always economic, political, and cultural issues involved in the practices of placing and distancing, displacing and replacing.

### *Distinctive neighbourhoods and social organization*

The idea of distinctive neighbourhoods is closely related to the concepts of placing and distancing (Madanipour, 2001). Richard Rogers had been working as the head of a task force to develop strategies for an Urban Renaissance in British cities (Urban Task Force, 1999). Their report, published in 1999, put forward a series of recommendations for urban development, to bring people and life back to the city. It envisaged cities as a neat system of neighbourhoods. Each neighbourhood was supposed to have a clear centre and a series of necessary ingredients. This idea of the neighbourhood as a distinct place with a clear centre, edges, and a mix of activities is also a theme of new urbanism, which had become a widespread idea in the United States from the 1980s onwards. This echoes the British new towns, like Washington and Milton Keynes, designed in the 1960s, which were a series of so-called villages around a transportation network (Llewlyn-Davies et al, 1966; Llewlyn-Davies, 1972). The two generations of British new towns in the second half of the twentieth century were mostly designed based on this idea of distinctive neighbourhoods, clearly separated from each other by some distance, to give the impression that each neighbourhood is a separate world on its own (Osborn and Whittick, 1963).

The idea is also reflected in Kevin Lynch's book, *Image of the City,* and his five elements of reading and designing the city, where he is advocating distinctive districts and clear edges (Lynch, 1960). Lynch looks for order, clarity, and distinction, trying to overcome the actual ambiguity and overlaps that are hallmarks of large cities, which are expressed in a diverse range of mental maps drawn by people from their memories. This desire for clarity and distinction partly goes back to the idea of neighbourhood units, which had become the urban design orthodoxy in the mid-twentieth century (Mumford, 1954). The idea was generating identifiable entities in the middle of a large and growing metropolis. Neighbourhood units were envisaged as a mechanism for social cohesion, focused on a school in the middle of an area, which would also be used as a community centre (LeGates and Stout, 2020). This practice of placing was accompanied by the practices of distancing, which were developed by land use planning. Around the turn of the twentieth century, mass migration to the United States had created fears of the incoming strangers in large cities like New York. The solution was devising a planning system that would classify people and activities into distinctive categories, allocate specific places to these categories, and ensure to keep apart what was thought to be incompatible (Cullingworth and Caves, 2013).

The idea also goes back to the British garden cities more than a century ago, which was also a response to the growth of the metropolis, but this time as an attempt to combine the city and the countryside, somehow trying to give shape and order to the process of suburbanization that had already started in British cities in the second half of the nineteenth century. What the garden cities movement wanted to create was an alternative to the city in the form of a small town. A form of micro-urbanism that would change the scale of the city from a metropolis to a village or a small town, made of identifiable neighbourhoods (Howard, 2003; Madanipour, 1996).

The context of this desire for distinction is the nineteenth century's twin processes of industrialization and urbanization, which created large cities full of people who were uprooted from their towns and villages, and had gathered in cities, living in miserable conditions, and without much connection with each other (Briggs, 1968). Ferdinand Tönnies (1957) was one of those analysts who distinguished between community and society. The old bonds of community and kin were broken, and the new bonds were at best functional and contractual. So, a major topic of concern becomes the alienation of the crowds, the disconnection of people from their cultural roots and from each other, and the social problems of isolation and lack of cohesion, which have remained to this day. The challenge becomes how to address this problem. One of the legacies of the nineteenth century in European cities was the institutions, such as churches, clubs, and unions, as well as spaces, like public parks, which were set up to confront this social fragmentation, and for the creation of new social bonds, a trend that has continued ever since (Figure 7.5).

*Figure 7.5* Marginalized neighbourhoods that suffered from placing and distancing are reconnected to the city by cable cars (Medellin, Colombia)

Reorganizing the city in the image of a series of villages is rooted in the romantic tradition of the eighteenth and nineteenth centuries, which nostalgically looked to the past, to the apparently peaceful life in rural areas, as an ideal. As mentioned earlier, Jean-Jacques Rousseau's ideal community is a village where everyone does everything, and there is no division of labour; it is imagined as a fully integrated life in complete contrast to the fragmented modern urban life. This idea of life in a village is still with us today. The majority of populations in English-speaking countries still prefer to live in a village, rather than a city, which is why suburbanization has continued with such force.

Garden cities, new towns, and new urbanism, suburbs as well as urban villages, and the more recent initiatives such as 15-minute cities and walkable neighbourhoods, are attempts at keeping this dream alive. It is this idealized image of life in the city as if living in a village that is evoked by the idea of planning and designing neighbourhoods. The idea that you can live an integrated life, with easy access to all your needs while being close to nature, even if you are in the middle of a world of high mobility, rapid change, social fragmentation, and alienation. The materialization of this dream is often limited to an image, unable to change the reality of living in a metropolis, where long-established land markets and social geographies resist any considerable change. Much has been said about the warmth of place, and the coldness and abstraction of space. A distinctive neighbourhood is expected to offer a place with an identity, escaping from the alienation of space, it is where authenticity is sought, but based on an image of a reality, rather than the reality itself, which is not ultimately desired in its authentic form, as it would imply abandoning the modern forms of living.

### *Places, clusters, and gaps*

A large urban area is usually a collection of clusters. For example, a map of London will show a cluster of government buildings, a cluster of places of entertainment, a cluster of educational institutions, a cluster of offices, a cluster of financial institutions, and so on. It will also show clusters of social and cultural groups, where higher and low-income groups and specific ethnicities are clustered. At the more detailed level, clusters of basic services have taken the form of high streets at the heart of city districts. The formation of these clusters is a historical process, with many decisions by individuals and organizations, as well as structures of government regulations and market mechanisms at work to generate this order of clusters. It is a process of placing, in which different functions and groups find spatial expression. In spatial economics, there has been longstanding interest in the reasons why competing businesses try to locate near each other. In the old Middle Eastern bazaars, this is clearly visible, where clusters of similar trades have created specific wings of the market, such as the jewellers in one street, spice traders in another, and so on.

Cities can be read through their clusters of functions, social hierarchy, and cultural diversity. They are different systems of distinction that find spatial expression, and once established, reproduce themselves. It is important to note that these are not natural processes, as some people have suggested. The Chicago School even used expressions from natural sciences to describe the emergence of social clusters, as if we are witnessing ecological processes at work. But the emergence of clusters are social and historical processes, rooted in particular circumstances and contexts. While individuals, households, and organizations may have a degree of choice in their location, they are operating within a context that provides opportunities and sets limitations. These limitations are highest when you are at the lowest ranks of the social hierarchy, where you are more likely to be placed, being given a tight spot in the social and spatial arrangements. But if you are at the top of the hierarchy, your circle of freedom to choose is much larger.

The practices of urban governance and urban design and planning tend to treat the existing clusters as given, trying to work with this grain. In Berlin, the city is divided into what are called Living Areas, which are identified on the basis of a series of socio-economic features of each area (Falahat and Madanipour, 2019). It is a move away from the past when the city was subdivided on the basis of its physical structure of roads. It shows more sensitivity towards social differences within the city so that the necessary responses can be made to people's needs. It identifies the socio-economic clusters and uses them as a tool of city administration. The idea of distinctive neighbourhoods becomes an urban management consideration. There is a transition from the emergence of clusters, to recognizing them, and to using them as an instrument. In what has been probably the largest wave of urban development in human history, Chinese residential areas in the past 40 years have been predominantly developed as gated compounds (Xu and Yang, 2009). This draws on the history of imperial China, on the communist era's integration of working and living space, on the structure of housing production by large companies, and on the fear of people from large waves of incoming strangers. Although recently this approach has been officially abandoned, and gated compounds are to be opened up for easing urban mobility, resistance by the local populations has not gone away.

Neighbourhoods are also a tool of production. The structure of the housing market in the UK is dominated by a few large developers. For these speculative developers, infill and single developments are not very profitable. Their pattern of work is to buy large parcels of land, borrow from the banks at low rates of interest, mobilize their sizable productive capacity, and develop large-scale residential areas. These operations are primarily driven by profit, and a captive market is not able to make substantive demands on the quality of the products. Economies of scale is an important part of their production of space. To sell their products, however, they need to promote them as a community, as a place, and as a desirable, prestigious destination.

The term village is frequently used, just to draw on the desire for rural living among the urban populations. As we saw in Newcastle's Great Park, these developers claim that we do not produce buildings, we produce homes and communities, make a place with all its positive connotations. However, the idea of a village does not have the same connotations in various parts of the world. In many countries, a village is a place of disadvantage and poverty, where agriculture is not rewarding, climate change is a threat for earning a living, and the rural village is a place to escape from. The urban village also has a negative image, as in China, where the urban village is a place of disadvantage, the place of ghost workers, unregistered and without access to services, where they can live in the middle of the city, but on the margins of society. The integration of existing villages into the growing metropolis has resulted in 'urban villages', which are spaces of ambiguity and alterity, where the normal rules do not apply and services are suspended for the unofficial residents of the city.

Placemaking has become the honorary title for a wide range of professionals. In the dichotomy of place and space, place is subjective, meaningful, and made by people who inhabit a space, making it their own over time. So how can the built environment professionals claim to make places, while making places can only be done by those who inhabit a place in the future, not by those who make it for an unknown market? In most housing development schemes, houses are built for anonymous, future buyers, who are yet to be found. Placemaking, like building communities or villages, becomes limited to promotion and branding, at best indicating an attempt to think holistically about a development.

For several years, a policy of neighbourhood planning has been at work in the UK. This is introduced as an initiative that gives the local populations the power to define the boundaries of their neighbourhood and plan for its future. It appears as a bottom-up, empowering tool to help people decide for their future. However, as it was introduced by a conservative government in favour of deregulation, it is viewed with suspicion. Such plans need local capacities to mobilize and to produce a plan, which is beyond the scope and ability of many low-income areas. Neighbourhood plans are not allowed to say what they don't want. Many middle-class areas don't want new housing in their area, but they are not allowed to say that in their plans. A plan is expected to say what people want for their area, but this requires investment, which comes from the private sector investors, who are interested in areas that offer good returns on their investment. So, there are development pressures in desirable areas, where investment is available, and where better-off people live, but they fight this pressure off, as they don't want new housing development that increases the density, traffic, and noise in their area. But this cannot be done through neighbourhood plans.

In less affluent areas, there is no appetite for investment, and no amount of neighbourhood planning can persuade private investors to invest in such areas. This is why the uptake of neighbourhood planning has not been very high.

When investors become interested in less affluent areas, they become the agents of displacement and gentrification. The area becomes attractive due to its location. Existing inhabitants are displaced by higher rents or forced out through redevelopment schemes. The *placing* process through which the market price mechanism, public policy, and social relations have produced a social geography is now being shaken. The economic foundations of that social geography have radically changed, the status, role, and strength of low-income workers have dwindled, and these workers are seen as being unduly residing on expensive land and property. While the mid-twentieth century had reorganized the built environment to accommodate the needs and demands of industrial workers, now the disorganized, internationalized, and poorly paid workers in the service sector are not considered as a threat, and thereby they are displaced from desirable locations and placed in the more marginal locations and conditions.

These structural economic changes have led to an increase in social inequality, as acknowledged by various organizations and analysts. The European Union, the OECD, and the UK government have all commissioned reports on how social inequality has risen in recent decades, marked by the decline of the welfare state, the emergence of neoliberalism, and the period of austerity that has followed the crisis of neoliberalism. With the rising tide of social inequality, the practices of placing and displacing are supported by the reinvigorated practices of distancing, removing the disadvantaged populations to farther distances on the periphery, or erecting barriers that would keep them apart.

The historical example of distancing has been suburbanization in some countries, leaving the poor in inner cities, and peripheralizing them in others, creating low-income banlieues. When social inequality grows, distancing becomes even more needed by the wealthy, and the barriers also grow, as in the gated neighbourhoods. When the two sides of social inequality are forced to live next to each other, distancing finds harsher expressions. In a regeneration scheme in Dublin, where the two social groups were separated from each other by a high wall, the children on the poor side of the wall called it the Berlin Wall. In another such scheme in Milan, the barriers that separate the two sides may be well designed, but the effect of placing and distancing would be the same. The barrier can be harsher still, as in cities where the threat of violence is used for distancing. Cities also practice un-placing. The growth of homelessness in the richest cities shows that the practices of placing and distancing have completely failed, leaving some people placeless and distance-less. They are close by and everywhere, nomadically fleeting and disobeying the rules of spatial order.

The creation of a vibrant place, where the image of a busy and happy city is played out, may be taken to be the expression of authenticity in urbanism. However, there are significant gaps between image and reality in urban development schemes, and the authenticity gap may not be filled for years before the area becomes a fully fledged part of a city. But the process of creation and transformation of the area may rely on placing, displacing, and distancing, where the apparent authenticity of a vibrant neighbourhood may stand on a dark platform of injustice.

## Looking for a place of origin

The reference point, on which the authenticity of something may be based, could be a place. Looking for an origin can take the search to mythical and imagined spaces, to the presumed purity and simplicity of natural places, to the home in which we grew up, and to the town that has always been there. This search can be found throughout history, both as an inspiration for reuniting with the lost origins and as a driving force for transforming the status quo. It has been used by reactionary forces to assert their rootedness and inclusion in a place, and thereby exclude others as outsiders who do not belong in that original place. It has also been used as a rallying cry for fighting injustice and domination. The modern world has given many of its inhabitants a degree of wealth and prosperity, which were previously unimaginable to be accessible at such a wide scale. This world has also given them a sense of homelessness, disconnection from their roots, and a permanent desire for reunification with themselves and their homes. The idea of authenticity is the name that is given to this search for the unity of being.

This unity, however, tends to remain as a pure desire: either the presumed origin was no more than a myth, or the person, the place, and the circumstances that frame them have changed. Place is one of the strongest elements in claims to authenticity, as it involves the continuous presence of a population in a particular location – claims that could disregard the many social and spatial changes over time that have transformed that continuity. Our individual memories and collective myths may be mere fantasies, or if they were true, they may not have survived the passage of time in their original state. As Heraclitus had said, you never step twice in the same river, and therefore a return to the same place is never possible; everything has moved on, and the gap can never be filled. Nevertheless, the drive for bridging that gap is a driving force of life, looking backward and forward at the same time. The desire for authenticity, as expressed in the particularity of places, as seeking distinction through association with those places, would need to be understood as a combined relation of similarity and difference, rather than simply an exercise in making distinctive narrative identities.

The practices that produce the urban space may claim to make places, drawing on the positive meanings of particularity versus the claim to universality of modernist practices. The image of a vibrant urban environment becomes the yardstick of authenticity in urbanism, without due attention to the diversity of conditions that make up urban life, and to the processes of placing and distancing that are at work to construct and reconstruct diversified and unequal social geographies. However, meaningful places are more than images; they are made by people through time, rather than instantly created by professionals. The apparently benevolent claims of placemaking may be aligned with the practices of placing and distancing, which tend to maintain and reproduce the existing inequalities.

# 8 Society

This chapter examines society as the possible point of reference for authenticity. It is the last in the series of possible origins that we have investigated so far. If the object is the city and the origin is society, can there be a relationship of authenticity between them? As mentioned in Chapter 4, in their analysis of the authentic works of art, and despite their differences, Benjamin and Adorno both shift their emphasis from the author to the object, so that they can avoid the cult of personality and instead analyse the artwork within a critical political framework. While Benjamin (1969) finds the authenticity of the work of art in its presence, its 'aura', Adorno links the authenticity of the artwork to its content (Adorno, 2002:45), and its extent of reflecting 'the self-unconscious historiography of their epoch' (Adorno, 2002:182). These concepts can be recognized in the idea of authentic heritage, which is based on providing empirical evidence that a present cultural item belongs to a specific period, place, or culture.

Beyond the works of art and items of cultural heritage, can these concepts be applicable to the city and its planned transformation? This is not to treat the city as a singular work of art but to consider whether the focus on the city as the object can open the analysis to an understanding of the social meanings of authenticity. If we extend these critical approaches by Benjamin and Adorno to the city and aim at interrogating the authenticity of a city, we will need to describe its aura, identify its contents, and examine its relationship with its epoch. This takes us to further questions: how can we describe a presence or an epoch in a city, and can a city be considered a work of art at all, or is this an unreasonable extension of a concept? This chapter engages with the presuppositions, associations, frameworks, particularities, and values that are involved in examining authenticity, as well as the potential capacity of authenticity to raise a critique and articulate an ideal.

## Association and fragmentation

If we take society as the point of reference on which the authenticity of something is based, we need to examine the notion of society, and whether it can be described in a way that would provide that point. The idea of society refers to an

DOI: 10.4324/9781003451778-8

aggregation of people who are related to each other in some arrangement (OED.com). In the English language, its earliest uses in the twelfth century indicated company, fellowship, and relationship. Later it described professional and religious groupings, such as monastic communities, and by the seventeenth century, it meant a community with shared customs, institutions, and laws (OED.com). This early sense of membership, association, and cohesion, which emerged in medieval urban society, was carried to the large urban agglomerations that grew after the Industrial Revolution. The new urban condition, with its atomization, fragmentation, inequality, and diversity continued to be described by a historical term that implied cohesion and association, leading to tensions and contradictions that reflected this gap between concept and reality (Madanipour, 2023).

In the notion of authenticity that draws on society as its point of reference, it is this cohesive meaning of society that is implied. Something is declared authentic by virtue of being linked to a specific society as its origin. However, as we have seen from other origins of authenticity, on closer inspection this concept can lose its coherence and clarity. With its complexity and diversity, society is not a single homogeneous entity, and therefore, it is difficult for it to serve as a single point of reference. This has been much debated in the discourses of identity, and how the identity of an entire society cannot be reduced to simple and narrow descriptions. The political and social implications of producing narrow descriptions for society have often led to exclusion and discrimination for those parts of the population who would not fit into these descriptions (Madanipour, 2020). If we are to avoid narrow, single descriptions for a complex phenomenon such as society, how can anything be called authentic by referencing such complexity? If there is an authenticity to be found through reference to society, it would have to take into account the diversity of society's constituent parts.

The legacy of the historical tension between the concept of a cohesive totality and the emergence of a fragmented multiplicity has led to distinctive ways of thinking about society, with direct relevance for the analysis of authenticity. With the increasing significance of the individual as the centre of the social world, one approach describes society as a collection of individuals, questioning the existence of society as an entity that is independent of these individuals. Society is made through the acts of individuals and the institutions and the processes that facilitate and support this group formation. This approach tends to adopt a nominalist position, which considers society an abstract term, a metaphysical concept, a mere name that does not signify any external reality (Rodriguez-Pereyra, 2019). Society is made of 'acting people' (Blumer, 1969:85), engaged in a process of interaction, without the need to resort to unknown and unknowable things or fixed structures and systems (Joas and Knöbl, 2009; Rock, 1979).

An opposite approach gives priority to society over the individuals. Society stands in sharp contrast to the individual, both constraining the individuals and supporting them through a range of institutions, organizations, and relationships (Buchanan, 2010). Rather than a mere aggregation of individuals, society describes social life in general and how it is organized in different units (Calhoun, 2002). Society exists as an entity that is independent of its members, exerting power over individuals but not depending on them (Durkheim, 1982). It is a holistic perspective that can be traced back to Hegel, giving priority to the social over the individual. As Hegel (1991:§239) argued, the (civil) society is 'a universal family' with duties and rights over the individuals: 'the individual becomes a son of civil society, which has as many claims upon him as he has rights in relation to it' (Hegel, 1991:§238).

The two approaches both make a distinction between individual and society, each placing its priority over one end of this dichotomy, with immense political, economic, and cultural implications. They are associated with a distinction between social determinism and self-determinism, which we have seen from the start of this book to be a key tension in the debates about authenticity. The range of views about society is of course much richer and far more complex than this simplified dichotomy, with a vast range of views that can be found within and between the two camps. The holist camp does not necessarily presuppose society to be a homogenous unity, and the individualist camp does not hold an image of society as a formless heap of sand. In both cases, there can be found nuances that try to reconcile elements of generality with particularity.

To go beyond a simplified dichotomy, the analysis of society needs to be dialectical rather than categorical, looking at how the different approaches have tried to make sense of the conditions of, and the relationship between, self and other, individual and society. It would also need to include an analysis of social ontology, that is, what society is made of. Beyond a multiplicity of individuals, different perspectives have emphasized social institutions, such as households and communities, but also the various organizations that structure the affairs of a society, such as its schools, courts, temples, etc. Society may be structured along social classes, which stratify society in a hierarchical way, and simultaneously made of diverse groups, along age, gender, ethnicity, culture, and physical ability. Furthermore, society often includes a territorial dimension, that is, it is located somewhere, thereby the significance of place, boundary, spatial subdivisions, and associated concepts such as nationalism and statehood as organizing principles. If the authenticity of something relies on society as its origin and point of reference, it is likely to end in these descriptors of society, each of which may stand for society but does not describe it as a whole.

An example would be the common practice of identifying national characteristics, or epochal styles in the arts. If you can prove that you have clearly defined a national character, you may be able to decide whether someone or something authentically belongs to that nation. But how can a nation, in its range of diversities, be captured by some descriptors that are not mere stereotypes? It would have to select some features from some people in some parts of a territory and extrapolate them to the entire nation. This selective identity then becomes the platform on which authenticity, as a form of identity, is evaluated, with inevitably limited validity. While stereotyping national characteristics is a more popular pastime, describing the epochal styles in the arts is a far more expert-driven exercise. From the vast range of cultural products that can be found from a period of history, some products, made by some artists, in some parts of some territories are selected as representatives of that epoch. Again, it is a selective process that determines the cultural identity of a period by narrowing it down, using one of the period's constituent parts to stand for it in its entirety. Subsequently, the authenticity of something is tested by its relevance to this narrow selection, leading to a limited outcome.

As many social analysts have argued, any interpretation of society would be shaped by the social position, as well as the disposition, of the analyst (Bourdieu, 2000). In social research, therefore, awareness of positionality becomes significant. This does not mean excessive relativism, that every account is ultimately a personal story among others. But it shows that in analysing society, we are faced with a multiplicity of perspectives that are narrating a multiplicity of forms and layers. There are no single interpretations of social structures or their constituent parts. If authenticity is a search for a reliable origin, which could explain something's uniqueness and distinctions, the cacophony of voices and the multiplicities of society do not seem to offer a single, clear pathway.

Therefore, the totality of a society cannot be described in a way that would offer a clear point of reference for authenticity. Instead, some elements from within a society are selected to represent that society and secure a coherent identity that aligns with its structures and processes. This exposes the process to a critique of representation and its associated processes. However, this critique does not mean a rejection of all claims as mere personal stories, but to examine each claim with an understanding of its potential limitations. Rather than searching for a single description of a complex society as its basis of authenticity, there may be different forms of authenticity that apply to different sections of society. This does not mean taking refuge in relativism when faced with the claims of authenticity, but the need for recognizing the diversity of elements and layers that make up a society, and how valuable traits may be found in each layer.

## Common frameworks

An important aspect of a society, where meanings such as authenticity may be found, is its culture. If we use 'culture' as the reference point for authenticity, we would be judging something to be authentic if it is proven to be rooted in a specific culture. Culture would serve as the origin and the point of reference on which the authenticity of something is based. This would put forward questions about what we mean by culture, and how the depth of rootedness in it can be demonstrated. The notion of culture refers to three different meanings: products of human endeavour, process of improvement, and the totality of a way of life (OED.com). However, as in 'society', an analysis of culture through these three meanings would reveal their multi-dimensionality and porosity, making it difficult to locate that reference point (Madanipour, 2016).

The first meaning of culture denotes the products of human endeavour. In the UK, the most common application of the term culture refers to cultural products, which, according to the Oxford English Dictionary, include 'the arts and other manifestations of human intellectual achievement regarded collectively' and 'a refined understanding of this' (OED.com). However, this poses several challenges. The first challenge is selecting what counts as human intellectual achievements. The tendency has been to emphasize the best works in arts and sciences, but such achievements may be found in all areas of human life.

A culture is enriched by a proliferation of the works of art, but as we saw in Chapter 4, it is not straightforward to decide which works are authentic. The authentication of the past works, in terms of testing their provenance and authorship, is different from the authenticity of contemporary works, where such authentication may not be needed. A significant challenge is about how a multiplicity of products can provide a foundation on which something's authenticity, or any other quality, can be evaluated. This would require selecting some works, or some aspects of some works, as the point of reference, a process that would be inherently reductive and potentially arbitrary. While the multiplication of artworks is a sign of cultural richness, inherently this multiplicity means a range of different qualities, as quantity does not equate quality.

Works of art may be grouped under the banner of their styles and schools, and authenticity might be judged by the relationship of a work to that style. For example, if a painting is shown to be an example of Impressionism, then it may be called an authentic Impressionist painting. If a building or an urban area is proven to be example of Modernism, then it would be judged to be authentic representations of modernist architecture and urbanism. The test of authenticity in the works of art becomes establishing a relationship between an object and a specific set of characteristics so that the object is judged to be truthfully representing those characteristics. The problem is setting boundaries that determine those characteristics, in this case, what constitutes Impressionism or Modernism, and whether a work belongs in this group. This selection would also exclude the works that do not demonstrate, partly or fully, those characteristics, thereby excluded from the test of authenticity.

As long argued, cultural products in market economies are often treated as commodities, prioritizing their exchange value, rather than their cultural and social worth. These products are made and promoted by a culture industry (Adorno, 1991), that is interested in generating spectacle (Debord, 1994), attention, and consumption (Madanipour, 2011). The primary point of reference for these products would be their market value, which cannot be used as a yardstick of authenticity. Any claim for authenticity would lead to an economic consideration and a potential or actual market transaction. Cultural products have a variety of dimensions, but if their primary dimension is the abstract medium of money, any claim to rootedness may appear to be too shallow. Another dimension is the role of cultural products in social positioning. As the production and consumption of these products has been shown to be associated with asserting social status (Bourdieu, 1984), claims to authenticity would lead to a performance, which would also limit the depth of rootedness that such a claim can reach. These linkages show that cultural products are not floating objects, detached from any context, but their contexts are complex and multi-dimensional. They are historical objects, even when they are very new.

Some artworks have lingered through ages, penetrating the different layers of society, becoming one of the pillars of a culture, and even celebrated in other cultures, as exemplified by the work of Leonardo da Vinci. The famous diagram of the Vitruvian man by Leonardo is often seen as one of the high points of the Renaissance, bringing arts and sciences together in the proportions of the human body, generating an image used as a major icon over the centuries. However, as the name indicates, it was inspired by the ancient Roman architect Vitruvius, who had written about the human body's proportions. Leonardo's diagram also had its roots and precedents in similar diagrams in his time, as different artists had attempted to depict the idea, albeit in less geometrical and aesthetic perfection. As the story of Leonardo's diagram shows, such works do not emerge in a void; they are part of history, before and after themselves, and therefore authenticity indicates a chain of connections, a historical unfolding, and overlapping interpretations, rather than a fixed and final address at which the search stops.

This can also be seen in everyday functional objects, such as furniture, clothing, and food items. An examination of moussaka, which has been widely considered the Greek national dish, shows that it came into being in a cookbook that was published around 100 years ago. The Greek chef and author, Nikolaos Tselementes, who has been crowned the godfather of Greek moussaka, aimed at purging the dish from its Ottoman elements and returning to its ancient Greek roots. However, the recipe that he provided and has been largely used ever since, albeit in different variations, appears to have been inspired by the French gratin (Economist, 2024). It proves that 'authenticity is an elusive concept in food' (Economist, 2024:77). Even after its widespread appeal around the world, the dish has been made in various new forms, including new ingredients and tastes. The search for purity in cultural products often leads to ambiguous points of origin, as traditions are always forged through a process of interactions between cultures. Authenticity is a dialectical process, rather than a categorical fixity, an unfolding temporality rather than a static timelessness.

The second meaning of culture refers to the process of improvement, which goes back to the concept of cultivation, where the word originates in Latin. It is based on the idea that, if agriculture can cultivate plants and animals, culture can do the same for humans, through education in the arts, sciences, and sports. The question for our investigation is: can culture as the process of education and cultivation provide a basis for authenticity? As the history of early modern Europe shows, much effort has been made over centuries to cultivate the population towards codes of civility, making them distinguishable from the presumed savagery of nature, and from the so-called uncivilized populations in other parts of the world (Thomas, 2020). Rousseau famously dichotomized the life of the savage with the life of the civilized urban populations. The ubiquitous dichotomy between nature and culture lies here. However, while modern society is based on the necessity of education, debates have long raged about the structure and content of education, its relationship with power, and whether it brings about homogenization and conformity, rather than authenticity.

As Raymond Williams (1981) argues, education is preparation for a particular social order. School education, as well as the various cultural institutions and processes of improvement and refinement, has attempted to control the so-called natural ways of doing things. The argument that the playground is a conspiracy against the freedom of the child is an example of resistance against this control (Ward, 1978). Many commentators have nostalgically written about the freedom of playing in the street, and how designated playgrounds are a limit to the ability of the child to be freely integrated into the city. The entire movement of taking refuge in the natural environment, which is partly manifested in suburbanization, reflects this tension between culture and authenticity. If authenticity is urging the unity of the person's behaviour with their inner reality, then as we saw in the last chapter, that real person is imagined to be in a natural state, rather than a cultivated and refined one. Therefore, the notion of culture as refinement and civility seems to be at odds with the notion of authenticity, which is specifically nonconformist and resistant to social norms and conventions.

A problem with this critique, however, is assuming that all cultivation is an artificial layer, a mere performance. It does not account for cultivation when it becomes an integral part of a person, developing habits that become second nature to a person and a population, as we saw in Chapter 5. Can the gap between nature and culture be visible here? If a population is brought up to be polite, are they being inauthentic and hypocritical, as they are assumed to hide their real intentions behind a veil of politeness? If a population is angry and aggressive, are they being authentic, as they reveal what they feel? But what if politeness has become so embedded in a culture that it is used as an integral part of its identity? And what if anger and aggressiveness found in a society is not a sign of its natural state, but a reflection of its current difficulties, which can be resolved by the change of its circumstances? Most importantly, what if the politeness and aggression that we attribute to an entire culture are only the features of some parts of this population, which we are now extrapolating through some form of stereotyping?

When it comes to the city, the process of improvement has long been a preoccupation of architects, city planners, and municipalities. For several centuries, cities have been the subject of improvement efforts, from the drive for making cities more beautiful (Laugier, 1977), more efficient (Giedion,

1967), more competitive (Morris, 1979), and brought back to life through redevelopment (Sert, 1944) and regeneration (Urban Task Force, 1999). However, the improvement process tends to change the character of a city or its parts, replacing the old structures, which are seen as outdated and obsolete, with new spaces that are thought to be aesthetically or functionally superior. In this process, much that might have been considered to be the authentic parts of the city, that is, those parts with which the character of the city had been associated for a long time, could be considered obsolete, destined for demolition or transformation. Locating authenticity in culture as the process of improvement may seem to generate a contradiction, as the city is an evolving multiplicity rather than a frozen singularity.

The third meaning of culture is a way of life, a broad, anthropological meaning that indicates 'the customs, civilization, and achievements of a particular time or people' (OED.com). This meaning overlaps with the notion of society and therefore exposed to similar critical questions. The assumptions about the homogeneity and unity of culture are no longer accepted, as any society is made of a range of cultures and subcultures associated with different social groups, activities, geographical regions, and ethnic diversities. What passes as the culture of a society or group may just be the selections and ways of life by a powerful minority, which has been imposed on others by force, and presented as the inclusive culture of everyone. Even so, that culture is never fixed and eternally the same, as it evolves and changes over time. The closer we look at a culture that appears homogenous and consistent, the more diverse it becomes, making it difficult to pinpoint a stable origin that would serve as a basis for testing the claims of authenticity.

UNESCO advocates the protection of cultural diversity, as it has been under threat by nationalist and authoritarian politics. In their declaration on the subject, cultural diversity is 'the common heritage of humanity' and 'as necessary for humankind as biodiversity is for nature'; it is 'embodied in the uniqueness and plurality of the identities of the groups and societies making up humankind' (UNESCO, 2002:62). As the nation states in much of the world are at odds with the cultural diversity in their national territories, UNESCO rightly asserts the necessity of the recognition and protection of this diversity. However, these cultures cannot be taken as black boxes with a fixed interior that should not change. For example, the treatment of women and children within a culture, or its attitude towards its neighbours, may not be easy to accept with the sensibilities of modern society. Authenticity may be claimed by relying on an unjust way of life, but it would not pass the test of equality and democratic acceptability.

## Partial authenticities

Can there be an authentic city in its entirety? If we use different criteria that we have listed so far, there is some possibility for calling a city authentic. If the point of origin is the named designer, who is credited with the design of the city, or a particular approach to design, which they represent, the city might be called an authentic city. This is exemplified in the case of Brasilia, which was designated as a World Heritage site by the UNESCO in 1987. This designation was made only around 30 years after its construction, so the criterion of authenticity did not need to be concerned about the accuracy of the documents that proved its link to a bygone age. According to UNESCO, 'Brasilia is a singular artistic achievement, a prime creation of the human genius, representing, on an urban scale, the living expression of the principles and ideals advanced by the Modernist Movement and effectively embodied in the Tropics through the urban and architectural planning of Lucio Costa and Oscar Niemeyer' (whc.unesco.org, Brasilia). Its designation, therefore, is justified by reference to 'human genius' and its representation of an approach, thereby emphasizing the role of the designers and their design philosophy.

UNESCO's judgement of the authenticity of the city was based on the reference to the city's designers and their school of architecture and urbanism (Figure 8.1).

> The authenticity of Brasilia is guaranteed through maintenance of its architecture, urban design, and landscapes, all of which represent a new approach to urban living, reaffirmed by Lucio Costa and Oscar Niemeyer on the basis of the Modernist Movement's principles for 20th century architecture and urbanism.
>
> (whc.unesco.org, Brasilia)

*Figure 8.1* The authenticity of the city is based on its representation of modernist architecture and urbanism (Brasilia, Brazil)

The key features of the plan (Pilot Project) and the buildings are then described to support this judgement. This shows authenticity as evidence of representation, displaying a link between a phenomenon (a new city) and a point of reference (two named designers, an approach to design).

There are no doubts about the political and aesthetic significance of Brasilia. It has been used as a symbol of Modernism, both in terms of the qualities of its physical structure and the idea of creating a completely new city. Like the city of Chandigarh, which was designed by Le Corbusier in India, or capital cities like Islamabad in Pakistan and Abuja in Nigeria, the notion of constructing a completely new city with the aid of modernist architecture and urbanism was a powerful trend in the twentieth century. Modernism was based on a fascination with the new, and these cities displayed that newness in the most spectacular way. At the same time, as these cities show, the same set of ideas, which had been formulated by Le Corbusier and the Modernist Movement, could find different forms when it was applied by different designers to different locations. The city of Brasilia's suitability to be designated a World Heritage site was partly based on its fidelity to this new approach in design.

However, Brasilia is more than the Pilot Plan. The city has witnessed many disputes about whether a capital city in a major country can be kept intact as if frozen in time. There are many pressures to keep the city as it was designed and built, and other pressures to open it up for new additions and the transformation of the existing structures. Meanwhile, the city grows in its different parts, informal settlements emerge, and the new areas do not follow the same aesthetic taste and functional logic that was deployed in the core city. What is considered authentic Brasilia is only the designed core, a part of a city rather than its entirety. Other cases of towns and cities that are designated as UNESCO World Heritage are often their historic cores or some elements of cities. An entire living city is too complex and multilayered to offer the possibility of capturing its authenticity.

A smaller city that has been wholly designated in 1995 as a world heritage site is Ferrara in the region of Emilia-Romagna in Italy. It represents the 'humanist concept of the "ideal city"', which 'came to life here in the neighbourhoods built from 1492 onwards by Biagio Rossetti according to the new principles of perspective' (whc.unesco.org, Ferrara). It is a project that 'marked the birth of modern town planning and influenced its subsequent development' (whc.unesco.org, Ferrara). Regarding its authenticity, the city and its surroundings represent the Renaissance and its approach to urban design. 'Ferrara, City of the Renaissance, and its Po Delta is a cultural landscape that is exceptionally well preserved and is authentic in its form and design, materials, setting, spirit, and feeling. The originality of the urban fabric of Ferrara, along with its Renaissance design and layout elements, makes it a clearly recognisable Renaissance city' (whc.unesco.org, Ferrara). The city has inevitably gone through many changes, but, 'Despite a long history of damage to the property, it retains a truthfulness and credibility with regard to its expression of Outstanding Universal Value' (whc.unesco.org, Ferrara).

The medieval parts of the city, which are visibly more vibrant than its Renaissance parts today, are not mentioned in the description of its authenticity (Figure 8.2). The city is considered authentic in its extensive Renaissance additions, but this authenticity is not found in those parts of the city that were built before or after this period. If, as we saw in Chapter 6, authenticity in a city is defined by its social dynamism, then the most authentic part of Ferrara may not be its Renaissance heritage, showing that there are different authenticities at work in a single city. Moreover, in both Brasilia and Ferrara, the considerations of authenticity are applied to the physical fabric, leaving aside the social life of the city which has obviously gone through many changes over decades and centuries. This shows that the notion of authenticity, as used by the UNESCO World Heritage process, only applies to the built environment in some parts of the designated cities. The city in its entirety, and its human environment, in part or as a whole, are not included, as the defined criteria of authenticity would not cover them.

*Figure 8.2* The medieval parts of the city are not included in its account of authenticity (Ferrara, Italy)

A case of world heritage that may appear to include the physical and social aspects of the city is Edinburgh. The capital of Scotland since the fifteenth century, the world heritage designation covers its medieval fortress and the eighteenth century New Town, which was an influential model of urban planning and design from the age of Scottish Enlightenment. According to UNESCO, 'The harmonious juxtaposition of these two contrasting historic areas, each with many important buildings, is what gives the city its unique character' (whc.unesco.org, Edinburgh). Regarding its authenticity, it is stated, 'The level of authenticity in Edinburgh is high. Individually the high-quality buildings of all dates have been conserved to a high standard and the layout of streets and squares maintain their intactness' (whc.unesco.org, Edinburgh). Beyond this primary concern with the buildings, the statement of authenticity goes on to say, 'The property also continues to retain its historic role as the administrative and cultural capital of Scotland, while remaining a vibrant economic centre' (whc.unesco.org, Edinburgh). Despite this reference to the administrative, cultural, and economic roles of the city, the focus remains on the character of the historic built environment, without engaging with the rest of the city, its other periods, and the complexities of its human environment (Figure 8.3).

As these three cases of Brasilia, Ferrara, and Edinburgh show, they are selected as a representative of an era, an approach, or an author. Despite

*Figure 8.3* The focus of interest in authentic heritage remains on the historic built environment (Edinburgh, UK)

the difficulties of representing an era, which runs the risk of reducing its complexities and diversities, it is possible to see how the concept of authenticity may be applied to a part of the built environment in a city. Specific features of specific parts from specific periods in a city are selected as evidence of authenticity, suggesting that authenticity in a city is always partial. As shown by Conzen (1960) in the morphological analysis of the town of Alnwick, Northumberland, a town is a historical creation, made of layers that are accumulated in different periods. This suggests that a town can have different degrees of authenticity in its different parts, some remaining intact and others going through transformation. The search for the authenticity of a town or city would therefore lead to a number of origins, rather than a singular one. In this case, we can speak of a city's authenticities. However, questions remain about the relationship between the built environment and the urban society, which is partly addressed in the development of a normative framework through which authenticity is defined and valued.

## Whose values?

At its core, authenticity is similar to a Platonic ideal, supposing an essence that lies behind an appearance, demanding the appearance to be united with the essence. It is not a recipe for the unity of two equals but for the subordination of one to the other. As such, it expresses an unequal power relation, in which something finds value through its association with something else.

The point of reference for authenticity might be some values. Something is called authentic if it represents those values, thereby being true to some principles and frameworks. A prime example is UNESCO's tests of authenticity based on ten criteria for Outstanding Universal Value, which are mainly about a cultural property's representation of the different cultural and environmental points of reference. These criteria list these reference points, which can be roughly summarized as (i) a 'human creative genius', (ii) interchange of human values, (iii) cultural tradition or civilization, (iv) a significant stage in human history, (v) interaction with the environment, (vi) events, ideas, and beliefs, (vii) natural phenomena, (viii) major stages of earth's history, (ix) significant ongoing ecological and biological processes, and (x) biological diversity (UNESCO, 2017, para 77). The focus is mainly on the tangible and intangible items that can provide a testimony to significant cultural, historical, and environmental conditions, now or, as it is mostly the case, in the past. The key value that is shared between these conditions is the value of acknowledging and protecting some cultural assets that are in danger of neglect and decline, cutting the lines of memory and identity in human society and reducing the richness of human culture. Proving the authenticity of some cultural properties confirms those properties' contribution to the values of cultural diversity and continuity (Figure 8.4).

We may dispute the inclusiveness and acceptability of these values. Whether there are any values that are universal across different cultures,

*Figure 8.4* Urban rituals contribute to the construction of the narratives of identity and cultural continuity (Seoul, South Korea)

whether something truly represents a culture, which parts of that culture are represented, and whether it should be kept intact or be subject to change are some of the questions that may be raised when faced with these values, which are declared to be universal across the world. Nevertheless, what they show is that the process of authentication is based on a normative platform, with known or unknown, explicit or implicit presuppositions, which should be always open to examination.

As the cases of world heritage designation show, authenticity relies on a set of values, which only find meaning within a context. The ways in which these values are maintained are also contextual. The protection of historical memories may have been a shared value between the Japanese and European cultures, but they used completely different approaches to ensure these values were maintained. For one culture, that value was maintained by keeping a historic building intact, while for the other this was achieved through renewal. This is why the Nara document, as we saw in Chapter 4, was produced to address some of these cultural differences. We are also faced with different sets of values within each society and culture, as the continuous disputes between high culture and popular culture show. Public funding for some arts has long been debated as to whether it supports a high art that is detached from society, reflecting the values and preferences of an elite.

Major tensions about values are those between economic necessity and cultural worth, between development and conservation, between human needs and the natural environment; dichotomies that reflect the diversity and at times incommensurability of different values. A famous dichotomy in values that Adam Smith noted, and Karl Marx emphasized, was that between the exchange value and use value. In these tensions, the idea of authenticity is sidelined, and a utilitarian perspective prevails, looking for the usefulness of an object, rather than its truthfulness to a value. The question that emerges is whether truthfulness to an instrumental value can be called authenticity. If a building is constructed for the pure purpose of profit maximization, which is common practice in capitalism, it has been true to a value. Can it be called an authentic building? If the point of reference is an interest, rather than a value, can it still be called authentic? If a cultural value is to be free from monetary considerations, can the artists be condemned if they try to sell their work in the market? If a cultural object is to be outside the market considerations, what are the principles and norms that determine its meaning and value?

One of the values of such designations has been a distinction between the original and the copy in cultural understanding. In contrast to these cases of authentic cultural properties, there are many examples of fake historical and natural environments. Recognizing the allure of the historical environments, or the attractions of other civilizations and cultures, new places are constructed that mimic those places. A completely new street in Wuhan, China, is built in European mixed historical styles, so that it may appear to have gone through many decades of change and transformation (Figure 8.5). The new town of Poundbury in England offers instant history, replicating a medieval town, with its winding streets and period buildings. A neighbourhood in Doha, Qatar, is built as a replica of Venice, with its canals, bridges, and building styles. A village in Sweden is completely built in the medieval style. In many cities, the old facades of buildings are kept but their interiors are demolished and redeveloped, creating a mismatch between the appearance and the structure, between the form and the contents. All these examples show how it is important to know about and to protect what is original and what is a copy.

Authenticity becomes a process of truthfully representing a point of reference, an origin. That point of reference may not be a homogenous and pure entity, but there is a fundamental difference between an original and a copy. Authenticity is a test of being true to what an item represents. It is an aesthetic and epistemic measure with which an ethical and cultural value is assessed. The process that a built environment was shaped, and whether that process could be ethically and culturally accepted today, is not included in the considerations. The conditions of human life in that environment, and whether and how those conditions can be improved are not part of the process. The broader context of that built environment, which in many cases falls within a growing urban area with all its challenges, cannot be addressed in such considerations. What is at stake is the accuracy and reliability of an

*Figure 8.5* A completely new street was built in European mixed historical styles (Wuhan, China)

aesthetic and epistemic link between an object before us and a point of reference that lies within or beyond it.

As the chapters in this book so far have shown, the selection of a point of reference is a normative decision, which reveals the values of the judges of authenticity. The source of originality could be a self, an author, an object, a place, a time, or an event, which shows different ways of looking at the world and different values and priorities adopted in the process. Rancière (2013) discusses the 'distribution of the sensible' to refer to the inequality of access to cultural expression and communication, and Bourdieu (1984) shows the hierarchical nature of aesthetic taste in society, both revealing that cultural values cannot simply be considered to be universal and free of bias and discrimination.

## Being true to the urban self?

If we go back to the five presuppositions that we saw at the start of Chapter 2, we can put those questions to the city more directly. In ontological terms, authenticity might mean looking for an urban essence, something that lies underneath the many appearances of the city. However, the city is a place of multiplicity, diversity, and stratification, and any attempt at finding a single essence would be reductionist. There are many ways in which such an essence has been envisaged.

In economic analysis, the city is defined through its economic base, its key economic products and relations that are the main source of its wealth. This could be an economic formation, a sector, a particular organization, or a particular set of relations, which prevail in the city. For some, the so-called creative class is envisaged to form the heart of the urban economy, for whom the rest of the citizens are classed as second order, providing services to the core group. In global competition between cities, the identity of the city is presented by slogans, and its presumed strengths are narrowed to what is marketable. The city's diverse population may be represented by some dominant cultural group, and its history and heritage by some periods and objects. In all these formulations, the representation of urban identity would involve a degree of reduction, with inevitably negative impact on those who are excluded from the description, as they are seen to be unfit and marginal. This applies to the city's physical structure, which is represented through its monumental places, and its social structure, which is represented by and through the social elites.

In the construction of a single narrative to describe the city's authentic and distinctive identity, its uniqueness is emphasized, and its multiplicity, diversity and change are downplayed. However, all cities are unique, and no two cities are the same; so on this account all cities would be authentic by virtue of their uniqueness. This uniqueness is not made of a fixed and unchanging essence, nor is it reducible to marketable slogans and major social and physical icons, but rather through a vast array of multiple, diverse, and changing parts. Multiplicity, diversity, and change are not formless, and there are trajectories, hierarchies, conflicts, and ruptures that shape and transform them. Authenticity is an assertion of particularity, so depending on what we select as the object, the theme, and the origin of that object, the search could produce many authenticities, rather than a singular authenticity, for any city.

The second presupposition is about differential values, giving higher status to one part of this ontological divide. But if this divide shows multiplicity rather than just a dichotomy, then we tend to select some elements of the city as a reference point for its authenticity. The selection is based on asking if something is from a certain period, if it is located in a particular place, if it is related to a certain group of people, if it is made by certain makers, if it refers to a particular way of doing things, if it is related to certain ideas and beliefs, all serving as a basis for evaluating that element as authentic. It is a process of selection and evaluation, which should be open to critical questioning: who is making that selection and what values are being promoted? When it comes to the works of art or heritage of the past, the point of reference could be made more specific, but in the diversity of the city, this becomes much more complicated.

From a functionalist perspective, the authenticity of something is largely determined by how well it serves its purpose or fits into the criteria that we have selected to evaluate it. As we saw in Chapter 4, a mass-produced chair can be called an authentic chair if we select its function as its prime identity. The same applies to the authenticity of a plan or a design, which may be judged by the extent to which it fit its purpose, as defined by a client. However, there are always questions about the processes by which the decisions were made to build or change some part of the city, the way planning and design were conducted, who it serves, and the impact of the plan on the urban population. Fitness for purpose is necessary, but it is part of a complex process of urban development that needs to be open to critical scrutiny. The same goes with the authenticity of a city as a whole, judged by how well it matches the general image that we have of a city. However, this narrowing of designs and cities to single, simplified descriptions hides their diversity, complexity, and their inherent inequalities and conflicts. This shows how something can simultaneously be called authentic and inauthentic.

The third presupposition is an epistemological principle: the possibility of knowing the self. Can we get to know the city and how? Comprehensive planning practices in the mid-twentieth century claimed to collect all the necessary data about a city and know it in its entirety. However, this knowledge always was partial, perspectival, and static. Even now with the availability of big data, what can be known is merely quantitative, limited in scope, and partial at best. The knowledge of the city depends on the scientific disciplines of the observers, each with its own topics, perspectives, and paradigms. When it comes to local knowledge, this diversity multiplies. Each individual citizen will have a different knowledge of the city, depending on their life trajectories and habits. If the self of the city, and the observers who wish to know it, are both multiple and diverse, then knowing the city does not generate a single narrative. We will have many different experiences, beliefs, and narratives about a city's visible conditions or hidden truths.

The fourth presupposition is about unifying the appearance and the essence, so that the true self can be disclosed, or pretensions be dropped. But is unification desirable or achievable when we come to the division of the city into public and private realms, habitual and spontaneous events, deep desires, and social behaviour? The life of the city is lived in a range of spaces that allow being with others and away from them, all necessary for managing privacy and sociability. Can we envisage uniting them into a single, undifferentiated space? The city of Amsterdam has been famous for its lack of curtains, to show that the households have nothing to hide from others. In different countries around the world, socialists and communitarians argued for the reduction or even abolition of private spaces. Modernist designers tried to overcome the separation of the inside and outside through large openings and moveable glass walls. Despite these attempts, the city keeps its private interiors, deep secrets, and dark corners, unwilling to unify the inside and the outside.

The fifth presupposition is about our ability to make this unity possible, to turn the city into its true self. Even if we could find that true self, and if unification were desirable, the power to make such a dramatic transformation may not be within reach. Planning and design are based on envisaging the possibility of an alternative, better arrangement for life in society. But the historical experiences of planning have shown that plans are not necessarily implemented, and when they are implemented, the consequences have not followed the predictions, or they have been rejected as unjust or unfit. We frequently speak of inclusive and participatory processes for knowledge generation and plan development, but they often remain partial. Moreover, as the powers of planners have dwindled in neoliberal economies, the ability to materialize these alternative visions has either been curtailed or twisted towards narrow interests.

Authenticity has been associated with democracy. In other words, democracy has been named as the condition of possibility for authenticity, thereby being authentic means contributing to democracy (Varga & Guignon, 2023). Being authentic would require collaboration and participation in society. Therefore, the source of authenticity is not in the past, but in the here and now, where things are being created. If this is done democratically, it is thought, then a society could become the source of authenticity. When our beliefs and behaviour are shaped by others, we are not free, and our actions will not be authentic, but if we are part of that process of shaping, we have been engaged in authentic, socially mediated self-determination. This authenticity has been created democratically, rather than individually. This is an ambition to bridge the gap between the individual and society, but it has to face the problem that a city or society is not reducible to this collective action, and there are always gaps and limits to such a claim to authenticity. It requires designating society as the point of reference for authenticity, but as we have seen, such a designation will face serious obstacles. Authenticity risks becoming a symbolic representation of a minority as the flag bearers of society.

This is clearly seen in the process of public participation in urban planning and design, which are seen as democratizing the process of decision-making in urban development and urban governance (Madanipour, 2012; Madanipour et al, 2017). If a place or an activity is created through participatory processes, it could become a point of reference on which a claim to authenticity may be made. Something would then be considered authentic if produced through deliberation and participation, thereby reflecting the needs and wishes of a particular group. It can be said to be truly reflective of a part of society, as the local population have been involved in the making of that place. In reaction to the top-down decision-making by a powerful minority, in which cities are transformed through reliance on functionalism and technology, or through profit maximization in real estate, public participation has been hailed as a democratization of the urban development process. It has become a necessary ingredient of urban planning and design. At the same time, participatory processes have not addressed the imbalances of power in the process, may not be inclusive, and be frequently used as an instrument of legitimation and justification, rather than being truly open to debate and change of direction. Any claim to such authenticity would be at best partial and limited.

When a group creates something, the internal dynamics of the group and how they arrive at their actions and constructions may be based on equality and respect, or on hierarchy and force. Either way, their creation may be called authentic by virtue of being undertaken by that group. If authenticity is the expression of an origin, and in this case, the origin is the group, then what they have made can be called their authentic creation. We can differentiate between these two methods of creation, and prefer one to the other, but are we facing two different authenticities? The same ambiguity may apply to the characteristics of a group. If the members of the group democratically decide to invade their neighbours and steal their land and assets, is that an authentic action, and are they an authentic community? This aggression is truly reflecting their characteristics, as they have been voluntarily and collectively engaged in this invasion. We can condemn this aggression on ethical grounds, but can it be doubted that it is an authentic act, undertaken by an authentic community, democratically deliberated and jointly executed? We saw in the discussion of psychoanalysis in Chapter 2 that airing deep desires can bridge the inside and outside. How far does a collective democratic process reflect the deep desires of a group?

## Authenticity as critique?

The idea of authenticity, as traced from Rousseau to Heidegger and Sartre, expresses a desire for individuality in the face of anonymity and social pressure. Conformity and alienation brought about in the modern urban society tended to ignore the human individual, turning everyone into a cog in a complex system. The city and the human being have both been imagined as mechanical systems, as exemplified in Le Corbusier's idea of the house as a machine to live in, and in the idea of a human mind as a computer. Encouraging people to think for themselves, as Kant had advocated, and to be themselves, as the Existentialists argued, was a cry for freedom in the face of convention and oppression. As Marx argued, the capitalist industrial society saw its members as quantities, rather than diverse qualities. Despite the modern emphasis on the individual, the mass society that both capitalism and socialism advocated undermined the individuality and the diversity of life trajectories. The quest for autonomy and authenticity offered a pathway for emancipation from social pressures, which ranged from the old religious obligations to the new functional and bureaucratic structures. As such, the concept of authenticity emerged as a critique of the social pressure for conformity and the conditions of anonymity and alienation. The aim was to offer a model of social recognition that is not solely based on similarity but one that acknowledges and appreciates difference. However, there are many similarities between human beings, and it is through a combination of similarity and difference that identities are forged, and people can live together in society.

Modern society has been criticized for its social fragmentation. The formation of this society was made possible through dismantling the earlier forms of agrarian life in towns and villages, breaking the social bonds that formed stable and cohesive communities. The urban population was no longer tied to the beliefs and values of these older communities, both enjoying freedom from these bonds and losing the protections and support that they provided. In time, this society produced its own structures of cohesion, through industrial working practices and solidarity movements, structures that were dismantled later during deindustrialization and neo-liberalization. The pursuit of individuality, which seemed like a cry for emancipation from the oppressive pressures in both rural and urban societies, was also a force for social fragmentation in the new urban conditions. Individualism became itself the subject of critique by those who aimed at reconstructing some frameworks for social cohesion. The drive for autonomy and authenticity was both a social critique and the subject of social critique, depending on the context of the criticism and the position from which it was made.

The ancient maxim Be True to Yourself is an ethical demand that condemned lies and deceitful performances. In modern society, where new codes of conduct were developed to make life in large urban agglomerations possible, politeness and performance played an increasingly important role. At times, these codes of life in crowded places could appear cold and detached, as Engels (1993) found in Victorian London's pavements. Urban life developed its codes of conduct, which differed according to class and culture, but were ultimately a device for creating and maintaining a particular social order. However, as the Baroque period shows, these performative models could be excessive in their elaboration and pretension, turning high society into an artificial construct, clothed in unwarranted ornamentation. The simplicity and proximity to nature that the romantics advocated since Rousseau, and the rejection of ornamentation that the functionalist modernists promoted, have been among the critical responses to this artificiality and hypocrisy.

In our time, digital technologies have facilitated a widening gap between the person and the performance. What is on display in the digital media could be completely made up, undermining the trust in written accounts, photographs, and videos, as discussed in Chapter 5. During the age of photography, what was on display in photos and videos were almost always a direct reflection of the event as it had happened. The images were edited and manipulated to tell a particular story, so they were not necessarily truthful and unbiased accounts. But an image was considered a piece of evidence, which could be used as a platform for different interpretations. In contrast, digital media can produce artificial images about events that never happened, with people who were never there. It can copy someone's voice in a narrative, or someone's face and body in fake situations. The demand for authenticity becomes more important than ever, as the authentication of the sounds, words, and images becomes ever more complicated.

A flaneur may have been an observer of the novelties of modern cities, but this is extended later to all forms of observation: from colourful magazines to television screens and to tourism. Now we are all digital flaneurs, going through thousands of web pages without necessarily needing anything or wanting to do anything. The gap between the observer and the observed started to grow for the nineteenth-century flaneurs, as they were not interested in directly engaging with anything, only in observing, tasting, and enjoying the atmosphere of the city. This is now the case in digital flaneurs, who experience the world through a screen and engage with it with a pleasure of passive observation, indicating a growing gap in which any sense of authenticity may be lost.

The demand for truthfulness is one thing, and the practices of telling truths or lies another. Authenticity can be presented as an ethical standard, but like other standards of social interaction, it may not be implemented. In authenticating the authorship of a work, the material of an object, and the age of a building, the identification and implementation of the standards may be somewhat possible through documentation and observation. But in personal beliefs and behaviour, being authentic may be demanded, but may not be even observable by the person who is the subject of scrutiny. It is here that the idea of authenticity may lose its critical force and turn into another pretension, a false claim about being truthful, and a license for narcissism. Authenticity remains as an ethical standard, which advocates truthfulness in personal life and social encounters, even if it remains elusive.

Authenticity as an expression of particularity can also be a form of political critique. When an oppressed group tries to develop and assert a sense of distinctive identity, it undertakes a form of resistance, a form of being different and therefore not subject to the same standards and treatments as others. The dominant group may justify its dominance by the discourses and practices that claim to be inclusive and universal, while the subaltern group needs to break that claim to universality through a claim to its own particularity. This can be seen in different social groups, in terms of gender, age, physical ability, class, ethnicity, and various subcultures, who use their sense of distinction as a weapon of resistance against the dominant forms of power. This can also be seen in national and international politics, whereby authentic identity becomes a means of struggle against majoritarian oppression, colonization, and globalization. However, as we saw in the discussion of national identity in Chapter 7, the claim for authenticity can offer an emancipatory potential but also carries the risk of rigidity and suppression of difference within that presumed authenticity.

In claiming authenticity, a social identity is developed that may reduce differences to fit into a coherent narrative. Authenticity offers a claim for the truthfulness of that narrative, such as a national or regional identity, by which a population can create a cohesive sense of itself and prove it to others that this is its true identity. Resorting to authenticity becomes a marker, a boundary, but one that can be oppressive to those who are inside and aggressive to the outside. Resistance to oppression may be mobilized by using authenticity as a political critique, but this critique can be double-edged, thereby requiring an internal openness that could question the claims of authenticity.

It has been suggested that 'The desire for authenticity is a desire for new foundations, but how does one know when one has created them?' (Akhavi, 1998:462). These foundations may never be found or always be questioned for their reliability and stability. Rather than looking for solid foundations or completely rejecting the claim to authenticity in an anti-foundational critique, a post-foundational response would offer the possibility of treating those origins with caution, a degree of flexibility, and contingency, always open to critique, questioning, and change. A dialectical approach would be needed to allow for interaction and evolution, going beyond a categorical one that seeks rigidity and permanence.

Authenticity can be an act of resistance against the dominating powers, an act of asserting dignity and singularity in the face of homogenizing forces of mass culture. It may be based on a fictional narrative of that singularity, but there are always elements of singularity in that fiction. The act of telling a story about our identity, and considering it to be unique, is the place of power, where we can assert a degree of power over our own life. Authenticity may be no more than a story, but it is a story that we have contributed to its writing and through the construction of this narrative we have been somewhat empowered. The belief in the ability to construct this story is a source of empowerment and agency, even if this authorship is severely constrained by the circumstances. The ability to make a choice within different options reflects that freedom, and not necessarily the ability to change the circumstances that are beyond our control. How to respond to events, and not how to control their occurrence, maybe the possible expression of that freedom.

## An integral part of social life

Is authenticity a critique of social conformity, raised from the outside and aimed at social change, or is it an integral part of the social arrangements? Authenticity has long been an undisputed ideal for some but rejected by others as a narcissistic claim. There are those, like Charles Taylor (1991), who seek a middle ground, arguing that authenticity should be kept and supported as an ethical ideal. Taylor (1991:11) acknowledges the arguments made by the critics and supporters of authenticity, suggesting that they are both right, but the way forward cannot just be about a trade-off between the two camps. For him modern society suffers from three malaises: individualism and the loss of meaning; disenchantment that is driven by efficiency and calculation of results; and the loss of freedom that is brought about by the institutions and structures of the industrial society and its technologies. Individualism and relativism have emptied the notion of authenticity from its constructive meanings, resulting in alienation from the public sphere and the loss of political control to the highly bureaucratic and centralized structures. Therefore, he sets out to rescue authenticity as a moral ideal by the force of rational argument, against scepticism, subjectivism, and determinism.

Rooted in the modern culture's rational individualism, and given an emotional character by the romantics, the history of authenticity since the eighteenth century shows the need to find a reliable foundation, external (nature, society), or internal (self). This link to an origin could offer some sense of security after the loss of certainty resulting from secularization, providing a basis for beliefs and behaviour, and a reliable means of social recognition, even when it is defined in an antisocial formulation. For Rousseau, what was needed was listening to the voice of nature within us; and for Herder, it was the belief that each person has an original way of being human (Taylor, 1991:28).

The appeal to this originality has been deeply embedded in the modern consciousness, becoming a powerful ideal. As Taylor argues,

> It accords crucial moral importance to a kind of contact with myself, with my own inner nature, which it sees as in danger of being lost, partly through the pressures towards outward conformity, but also because in taking an instrumental stance to myself, I may have lost the capacity to listen to this inner voice.
>
> (p.29)

It appeals to 'my own originality' through which I define myself and realize my potential but also leads to a culture of authenticity that includes some of it 'most degraded, absurd, or trivialized forms' (p.29). To confront this individualism, Taylor argues that the general feature of human life is its dialogical character, which is manifested in language, and in how we develop our sense of identity through interaction with significant others (p.33). Thereby, he argues, any attempt at self-fulfilment that is detached from others is doomed to fail.

As an optimist, Taylor sees authenticity as an ethical ideal, as it can lead to self-fulfilment, to become the best you can be, but this freedom can be used in a responsible or an egoistic way, and there is always a struggle to choose the right path, opening the field of ethics and politics. Taylor's solution for overcoming subjectivism and relativism is to argue for shared horizons of significance, which establish what matters, and in which individual choices find meaning. Authenticity is linked to normative models of how society should be organized (p.44). His Hegelian analysis shows the way: 'Authenticity is not the enemy of demands that emanate from beyond the self; it supposes such demands' (p.41). In other words, as Hegel had argued, the ethical life of a community (*Sittlichkeit*) is the framework that would enable, rather than stifle, that sense of originality and freedom. However, the question is how far such shared horizons can be found, or how easily can they be (re)constructed, when the forces of social fragmentation are so strong, and the forces of social cohesion could slide into the reactionary rejection of others.

To fill the gap between a desire for social cohesion and the reality of social fragmentation, a communitarian line of ideas and practices has stressed the need for enhancing and reconstituting communities as frameworks for social cohesion (Bell, 2024; Etzioni, 1995; Tönnies, 1957; Walzer, 2021). It is often based on a critique of individualism and social fragmentation in the modern world, memories of the cohesive communities of the past, and the possibility and desirability of creating new communities. Frequently, this attempt at community building takes on a spatial form in the design and development of neighbourhoods and small towns, from the idea of neighbourhood units a century ago to new towns, new urbanism, and various forms of neighbourhood planning and design (Falahat and Madanipour, 2019; LeGates and Stout, 2020; Madanipour, 2001). At the same time, the other face of inclusion and cohesion that is associated with micro-urbanism is the possibility of introversion, social exclusion, and rejection of others (Blakely and Snyder, 1999; Madanipour, 1996; 2020). Therefore, if the relations of association and cohesion are taken as a reference point for authenticity, in which something is authentic if created by a community, takes us back to the tensions between a narrow sense of identity and an open one, between the categorical and dialectic processes of identification.

Authenticity is a form of narrative identity. However, while the critique of detached individualism is a valid one, as the excesses of the culture of authenticity are everywhere on display, the problem with the traditional dialogical model, which is also used in other Hegelian-inspired analyses of intersubjectivity and communication (e.g., Habermas, 1984), is that it is based on microsocial interactions, unable to account for the diversity and complexity of the modern urban society. In these societies, ties with others are often indirect, mediated through the abstract mediums of money, status, bureaucracy, technology, and mythology. Therefore, identity is formed dialogically, but via abstract others, rather than through the medium of significant others who are personally known. Taylor's communitarian thinking aspires to the interpersonal, dialogical model of social relations, which could provide authenticity with a moral force, but the reality of society shows how this has been eroded by high-density urban living, intensified mobility, and the advent of the libertarian digital technologies.

As an ideal, authenticity is based on finding an individual, distinctive pathway. How can a multiplicity of people with their different, individual pathways find a common horizon, in which they share their belief in their own individual authenticity? This looks like a contradiction, in which the individual's identity is defined in contradistinction from society, while always being a member of that society. It suggests being simultaneously inside and outside society, with all the tensions and contradictions that this implies. We are therefore faced with a major problem. The ideal of authenticity encourages people to be themselves, rejecting conformity and convention, and embracing creativity and innovation. But this poses the problem of life in urban environments, where living together requires common standards that would reconcile these different pathways, enabling them to live peacefully together. If they all reject commonalities with others and instead engage in individual trajectories, how can society survive? Two answers might be found to this challenge of social fragmentation. First, that the level of individual detachment from others is exaggerated, and life in urban society would always involve a high degree of commonality. Second, that authenticity is already one of those common standards, through which social recognition is made possible.

We explored the first answer in Chapter 2, where the idea of the self is always related to others, and therefore even solitude is not without the presence of others. Being raised in a language and culture is sufficient to inscribe the significance of the others, known and unknown, in psyche and behaviour. The second answer is based on the idea that society is largely structured by the powerful forces that shape the working and living conditions, where the capitalist economic configuration has emerged as an organizing principle that allocates a position to everyone in society, attaching behavioural routines and monetary value to what they do. In parallel, national laws and social conventions shape and frame behaviour in society. These political, cultural, and economic frameworks limit the freedom of choice, expecting and forcing individuals to behave in certain ways. Some choices, such as the consumption of goods and services, may appear to offer a degree of freedom, but these are also contained within, and driven by political, economic, and cultural frameworks.

In response, the only pathway to freedom seems to be inwards, as the only place which may appear to be under the control of the individual. It is here that people can exert some control over themselves, through an aesthetic attitude that imagines an alternative mode of life. What is significant, however, is that this imagined freedom is not an alternative to the strictures of social life but an integral part of it, making it tolerable by providing some space for manoeuvre. It is the other side of the coin, an integral part of the social structure, rather than an entirely different way of life. It is not a critique of society that needs to be raised, as Heidegger and Sartre thought, but it is an inbuilt characteristic of the modern urban society, carving a space of subjectivity through which life in the structured society would become possible.

In the subdivision of society into the individual and collective fields, authenticity, like religion in secular society, is restricted to a personal domain, detached from the social roles that are subject to political rules, economic structures, and social conventions. The search for authenticity becomes a personal quest, resulting from the fragmentation of the social world into functional spheres. Authenticity, as being true to self, remains a personal endeavour, at best manifested in intersubjective encounters, in pursuit of social recognition. Beyond the individual level, authenticity may lose its ontological significance, while finding an epistemological role in the process of authentication, whereby it is used as an instrument of examining the truth of expressions and objects. It distinguishes a first-person, qualitative concern for self-discovery from a third-person process of judgement, distinguishing authenticity from authentication.

This is rooted in the historical tensions of modern society, whereby romanticism emerged in the nineteenth century in response to eighteenth-century rationalism, but neither movement could push the other out of the field. They become the two interdependent pillars of society, intertwining instrumental calculation and affective engagement, science and art, and political economy and culture. They become different ways of thinking about the world and describing it, but none could knock the other out of the arena. The desire for authenticity emerges in this context, which shows how it is an integral part of modern society, where the different forces meet and where subjectivity is constructed.

However, if social structures allocate specific places to individuals, do these individuals accept their places passively and obediently, or do they resist it through various means, including the notion of authenticity, which is an attempt to bridge the introspective and expressive parts of life and thereby overcome social alienation? Resistance to conformity becomes an expression of freedom, but simultaneously it risks becoming a form of conformity itself, whereby social recognition is sought by performing social nonconformity. Standing out as unique and original becomes an application for the membership of a club in which all members have a similar claim. Authenticity as the ideal of self-determination and self-unification that overcomes social alienation is at once a critique of the social structure and an integral part of it. It is simultaneously a cause of, and a remedy for, this alienation. The pursuit of authenticity is an integral part of social identity, which includes both difference and similarity, both how you wish to distinguish yourself from others, and in doing so become similar to those with whom you wish to be associated.

## Authenticity and urban society

Authenticity is a concept that is often used for individual people, as well as objects, practices, and places. When it comes to cities, which are complex and multilayered agglomerations of people, other life forms, and objects, we are faced with the difficulty of applying the concept. It is a concept that is developed for a singularity but is now expected to be applicable to a multiplicity. There are some ways in which singularities are connected to form collective authenticities. This can be seen in the collections of buildings that make up a part of the city, the groups of people who work together to change their cities, the use of authenticity as a social critique and an ethical ideal. Overall, however, we may be unable to apply the concept to the whole city but can only do so in some parts of the city. Rather than finding an authentic city, we may find authenticities within cities, with different subjects, in different areas, and with different characters, all created through the similarities and practices that may link their constituent elements, based on selected conventions and values. As the themes and origins of claims to authenticity are diverse, and as the processes of linking the fragmented multiplicities of the city are not inclusive, the concept of authenticity may not find a single origin for a city as a whole, the processes of group formation, and the complex processes of urban design and development. What seems to be common across these different pursuits of authenticity and truthfulness is how it is an integral part of living in complex, tightly structured society, carving a sphere of life that is necessary for survival. The structured and the authentic are the two sides of the same coin, rather than one emerging as an outside reaction to the other, displaying a dialectic of particularity and universality.

Authenticity is a pursuit rather than a status, a process rather than a destination, and a dialogue rather than a declaration. It is a quest for particularity and singularity in response to the forces of universality and conformity, but particularity and universality are intertwined, together constituting a social order. Authenticity is not a static formula but a process of authentication, in which an advantage is claimed for one thing through proving its relationship with another thing. This is why any claim to authenticity should go through continuous critical engagement, rather than accepting its refuge in fixed foundations.

# 9 Authenticity and urbanism

Authenticity is a key concept in contemporary culture, as evidenced by its selection as the word of the year 2023 by Merriam-Webster, its continued importance for UNESCO in designating the world heritage sites, its controversial nature as the battleground between the major philosophical schools of existentialism and critical theory, and a range of ongoing political, ethical, and aesthetic debates, including the concerns about truth and trust in the digital culture and in political institutions. This book has set out to investigate the meaning of the concept of authenticity and explore its relevance for urban socio-spatial processes.

## The concept of authenticity

The common meanings of authentic refer to a state of being true, actual, real, not false, trustworthy, unique, and original. It refers to the singularity and particularity of something, against the uniformity and universality that are the hallmarks of modern society. It also refers to the truthfulness of something, against the false appearances and pretensive behaviours that are commonly found in complex urban society. Accordingly, I have defined authenticity as the evaluative relationship between an object and an origin. The designated origin is presumed to serve as the truth of the object, thereby endowing it with a higher value. The object may be animate or inanimate, singular or multiple, idea or practice.

Authenticity is a process of identity formation, in which disparate physical, affective, and discursive fragments are brought together under a single narrative. First, different fragments are grouped into two locales, the object and the origin, and then these two are related through a direct narrative line. In other words, authenticity is a process of gathering and meaning-making. The two concepts, the object and origin, are themselves multiple and fragmented; authenticity is a linear, narrative identity that gathers these fragments into two identifiable concepts (object and origin), and links them to construct a meaningful story. Rather than a quantitative effort at measuring that relationship, it is a qualitative account of valorization.

The relationship between object and origin has ontological (being and becoming authentic) and epistemological (authentication) dimensions, indicating

DOI: 10.4324/9781003451778-9

whether something is authentic, and how it is possible to demonstrate that it is so. The core of both the ontological and epistemological dimensions is formed by concerns about social recognition, anxious to establish the relationship between an object and an origin, thereby showing the worth of an object through its positive relationship with something else that has a higher value. Therefore, the claim to authenticity is based on authentication, which is a valorization process, whereby one side (object) is imbued with a social value (social recognition for being truthful) through its linkage with the other side (origin).

The ontological and epistemological dimensions of this relationship are closely related, as objects are judged to be authentic through authentication, a process that proves or disproves the relationship between object and origin. In authentication, the central point is proving the existence of a link between the two sides of a relationship, through which one side finds value by being truthfully associated with the other side. The examination of this relationship is an evaluation and a judgement, which is framed by social conventions, institutional settings, and cultural contexts. The examination of the relationship takes place through correspondence or disclosure, that is, showing how accurately and reliably the object and origin are related, or how the object becomes one with the origin through a process of unfolding, in which the origin reveals itself. As I have argued, it is possible to see authentication through two perspectives: a third-person process that looks for some evidence to prove or disprove that link, and a first-person process that explores, discovers, and discloses that link. While the claims of the third-person perspective are open for scrutiny by others, the claims of the first-person perspective are not.

When looking from the outside at the relationship between something and an origin, that relationship can be observed, investigated, and evaluated. This is a process of certification, establishing the truth or otherwise of the link between the two. Whether a text is written by a certain author, an object is made of gold, a building belongs to the Roman period, or a food is from Japan are all examples of claims to authenticity that can be tested and evaluated. However, when at least one of these two parts of the equation, that is, between something and an origin, is not visible or easily accessible, such as the mental states of a person, authenticity becomes a process of looking from the inside, from the perspective of the person who is experiencing that relationship. In this introspective form, authenticity is an attempt at self-discovery, building self-confidence, and a claim for social recognition, and as such it would be largely difficult to be judged by others. Only by closely knowing that person for some time can anyone test the accuracy and reliability of these claims for personal authenticity. The idea of authenticity as the disclosure of the inner self assumes a degree of homogeneity and stability for that inner self, but this self is shown to be fragmented, multidimensional, and transforming, to the extent that even the person who makes the claim may be unable to verify that stability.

Authenticity can be defined as a double negative: as the refusal of false identification. When defined in this way, it can potentially become a powerful force. No matter how fragmented your ontology is, when you are faced with what seems to be a false epistemology, you develop a sense of the self that becomes a driving force. A coherent narrative of the self is generated as a mechanism of self-preservation, a mechanism that is inbuilt in all life forms. It unleashes energies that may have both peaceful and aggressive faces. When you are described in terms that you find them to be wrong or demeaning, you wish to assert your identity as you see it, rather than the way that others set for you. This can be at a personal or group level. If a particular minority group is demonized in public discourse, it can lead to violence against that group. The negative narrative of identity is a precursor to increasing pressure on that minority group. The group may fight back, engaging in negating this negation, protesting that those descriptions are not true. This is a clamour for authenticity as truthful representation of identity.

What this group considers its truthful representation may be itself based on some false stories. But it attempts to generate a more agreeable story about the group. The stories against and for a minority group may both have elements of truth and falsehood, but what is the core of the tension is the question of power. The larger group's demonizing of a smaller group is a power imbalance that leads to injustice, and to authenticity as a quest for justice. The narrative of authenticity would be a negation of negation, as a platform for fighting injustice. The narratives against and for a minority are often intertwined with personal impressions and exaggerated interpretations of single cases. It is important that these false stories are unpacked, and their contents be critically examined for accuracy and truthfulness. This would be part of recognizing the power imbalance and the need for addressing it through acknowledging the claim to authenticity while recognizing its limits. The narrative of authenticity emerges to fill the gap between the third-person and first-person views of the self or the group. If others see us in ways that we think are discriminatory, or at least not adequately accurate in their description, we set out to correct that misunderstanding by self-descriptions that satisfy us. Authenticity is the bridge between the view from the outside and the view from the inside. By relying on the inside, it challenges the view from the outside, a process of negating a negation.

Therefore, authenticity may be interpreted in two different ways: representational/evidential and experiential/existential. In the representational/evidential meaning, authenticity refers to the process of evaluating and judging the accuracy and reliability of the relationship between something and a point of reference. If something can be reliably related to a particular person, material, time, place, and society, the observers may be convinced that it is an authentic representation of that reference point. In the experiential/existential mode, authenticity refers to the disclosure of an origin, which can be experienced by the person or group, but not necessarily by others. It can generate an aura in an object, a feeling of confidence, agency, and truthfulness in a person or group, but the observers may not feel the same or necessarily agree with these claims, which are not easy to communicate and difficult to observe. Both versions have an ethic of truthfulness, but the representational mode is more verifiable and falsifiable than the experiential mode, which cannot be put to the test of observation and examination. One is epistemic authenticity that depends on authentication, the other is aesthetic authenticity that relies on self-disclosure.

A key question is about the choice of the point of reference and the criteria with which the relationship is evaluated. If the origin is a detached self or collective belief, the meaning and character of the resulting authenticity may be widely different. Even in the case of a single self, the desire for social recognition and assertion of social status cannot be simply dismissed as nonsensical: in a world dominated by pressures that are beyond our control, descending from processes that we may find opaque and unjust, the desire for clarity of meaning and a degree of choice in our lives cannot just be wished away. But this desire can only be satisfied through intersubjectivity, beyond the private sphere and through exchange with others. Therefore, in both versions of authenticity, representational and experiential, epistemic and aesthetic, the struggle for status and recognition is the primary drive. Authenticity becomes a search for social affirmation, a process that is triggered by a claim to sincerity and truthfulness.

The notion of authenticity is based on the presupposition of inauthenticity. Certain objects, ideas, behaviours, people, relations, and conditions are thought to be inauthentic, that is, fake, artificial, and untrustworthy. In response, the notion of authenticity is presented as a search for reliable and truthful things. Without presupposing that things are inauthentic, as a critique of the status quo, there would be no search for authenticity as a positive way out of that undesirable state. This is rooted in a critique of society, whereby a lack of trust in people and things has prevailed, and where authenticity appears to offer a pathway towards truthfulness. We may assume that authenticity was a pure state before being spoiled by inauthenticity, but in fact we are first formulating a critique of what appears to us as inauthenticity; thereby it is inauthenticity that comes before a quest for authenticity. As such, authenticity is a relational and dialectical concept, only making sense in relation to its negation, inauthenticity.

Therefore, authenticity and inauthenticity constitute a dialectical relation, rather than a categorical dichotomy. This dialectical relation is not Hegelian, in the sense of necessarily leading to more advanced, positive outcomes, but in the sense that it is an open-ended interactive process that could go in any direction. The two concepts are necessary for each other, as one would not make sense without the other. We cannot think of authenticity without simultaneously thinking about what is not authentic, that is, inauthenticity. As it is a value judgement, it is based on a distinction that is described through a dichotomy. The problem with this dichotomy, which treats authenticity and inauthenticity as categorical states, is to think of them as fixed and unchanging conditions that are at odds with each other. Heidegger defines authenticity as being free from the 'they', but even if we try to avoid being a powerless pawn that follows the crowd, at a deeper level we are intertwined with the 'they'. We can never become completely detached from the society and culture in which we were brought up and live. This would mean that a dichotomy between authenticity and inauthenticity, between self and other, cannot be maintained at the most fundamental level; it is always an evolving relation, which takes different shapes and could be reinterpreted as authentic or inauthentic.

I have argued that examining the relationship between the object and origin requires analysing both the object and the origin, rather than taking them for granted as solid grounds. In other words, I have tried to show how these origins are based on selective narratives, which are context- and observer-dependant. It is through the social and historical contexts that something finds value for serving as a point of reference for other things. In the chapters of the book, I have identified and critically examined a series of origins for authenticity, and in doing so have tried both to develop a better understanding of the concept and to explore its implications for the city as the object under investigation. In this investigation, the object has been the urban society and space, and the origin has been analysed in seven categories of self, author, substance, event, time, space, and society.

The object may be in front of us, concrete and with a physical presence and accessible for description. The origin, however, is not readily accessible, as it is located somewhere else: inside the object, in the recesses of the human mind, and hidden interiors of the social world, or outside the object in events, times, places, and cultures. The origin is not given, it must be selected, and therefore it is a choice, the outcome of social and historical conventions and processes. It needs to be chosen, discovered, disclosed, and claimed as the truth of our concrete object, describing its identity, and proving its authenticity. But all objects are continually changing, with many parts and characteristics, each of which may have a variety of possible origins, and those who judge the claim to authenticity may have different opinions or may change their own views. Therefore, the search for an origin could be a never-ending process of discovering multiple and changing patterns of authenticity and inauthenticity for the same object. The way out of what looks like an infinite labyrinth may be to acknowledge that this is not discovering a fixed and final truth about an object but producing a possible description for it. It is a context-specific and observer-dependent value judgement, based on socially mediated experience, in a process of selection and identification, constructing a narrative that links two items, an object to a point of reference that might validate it.

The most important feature of this link is its claim to be truthful. Being authentic demands a unification of the outside and inside, of appearances and essences, removing the barriers to transparency and encouraging the disclosure of what lies inside. However, the claim to the consistency of the essence and the purity of the inside can lead to dogmatic outcomes with dangerous consequences for some groups, being rejected as impure outsiders. It can lead to self-seeking behaviour that disregards others as inauthentic, potentially detaching all objects and relations from their meaningful contexts and turning them into commodities for sale. This is why any claim to authenticity needs to be critically examined for its truthfulness, as well as its intended and unintended consequences.

The study of authenticity reveals a deep and ongoing ideological battle between those who embrace the idea of authenticity as an expression of self-knowledge, self-determination, and individual liberty, and those who argue for the social determination of belief and behaviour, rejecting authenticity as individualist, antisocial, and regressive. It reflects the long-held dichotomy between individual and society, which finds political, ethical, and aesthetic expression in the idea of authenticity. To avoid being trapped in this dichotomy, and looking for the creative potentials of the idea, I have investigated authenticity as a search for truthfulness, examining and problematizing what this process entails. In the context of an alienating urban world, in which powerful, impersonal forces dominate, the pursuit of authenticity reflects the need for developing self-confidence and social recognition, in a journey that is filled with unbridgeable gaps, colourful mirages, baseless claims, and exclusionary actions.

The longing for authenticity is a desire for resistance against the straitjacket of conformity, breaking the codes of familiarity and conventionality, which have been transformed into impersonal codes in high-density urban society. However, this desire for distinctiveness, when encouraged and embraced by many, becomes another common pattern, a vehicle of social conformity through a new code that would create similarity rather than the difference that was desired. Meanwhile, some of the key structuring forces, such as the push for commodification of all aspects of life, remain in place, which turn the gestures of nonconformity into market opportunity. These gestures may also be pacified as safe, personal aesthetic choices, directing the social pressures inwards into the private domain and acting as corrective movements within an overall system of dominance. Nevertheless, the creative potential of the desire for truthfulness can find a social force that goes beyond its conditions of possibility, triggering the emergence of new circumstances.

Authenticity is a social yardstick, an ethical standard, which is used to judge the truthfulness of a relationship between an appearance and an essence. It is an indicator of a gap between the appearance and essence, and the need for certainty about the reliability of that appearance. It is a link between the phenomenon and what is assumed to be its source of identification. It is used in making a claim to being authentic, in evaluating objects for their authenticity, and in demanding others to be authentic. It is therefore not limited to self-serving claims, but an integral part of social interactions. The presumed essence may not even exist, it may not be stable or accessible, the methods of testing the truthfulness of the relationship may not be reliable, and the outcome of the judgement may be debatable; nevertheless, authenticity remains a social value, a currency that plays a part in regulating social relations. By resorting to an origin, and claiming to be armed with truth, authenticity is a quest for empowerment in a world dominated by powerful forces that frame and constrain individuals. The origins may be multiple and shaky, and the claim to truth may be difficult to prove, so authenticity always remains a quest, a socially mediated process rather than an atemporal and aspatial destination. In examining the provisional nature of these origins, I have undertaken a non-essentialist, post-foundational analysis of authenticity that looks for its creative potential while being aware of its severe limitations and concerned about its problematic consequences.

Authenticity is based on a process of authentication, seeking an origin, a solid ground when feeling vulnerable in the world of contingency, a ground that provides distinctiveness and particularity amid multiplicity and universality, and freedom from conformity and constraint. That solid ground is sought in the self, in the act of making, in the substance of objects, in transformative events, in particular times and places, and in society and culture. The ground, however, is never unitary and solid, and authentication is always a process of selecting, bridging, and constructing, rather than treading along a well-paved path towards a well-signed, fixed destination. In urbanism, as in the personal and social pursuit of authenticity, this does not mean giving up the search for a solid ground, but to acknowledge its social basis with its potentialities and limitations. Rather than withdrawal and disillusionment, this opens the fields of possibility, as it releases the action from rigid origins and allows it to explore new horizons through interaction with others in the public realm.

## Authenticity and the city

The primary reference point for authenticity is the self, both in knowing it and in making it. As such, the maxim 'be true to your self' is a key concept in the discourse of authenticity. It implies five presuppositions that correspond to the main meanings of authenticity: ontological split, in which the inner and outer self are not the same; differential value, where the inner layer is given priority; epistemological possibility, where knowing this inner self is possible; normative desirability, where the unity of the inner and outer layers is advocated; and the power principle, which assumes that you have the power to cause this unification. However, these presuppositions and this recipe for the unification of the inside and the outside, mental processes and bodily behaviours, essences and appearances, private characteristics and public performances, should be critically examined.

Rather than concluding that selecting the self as an origin does not provide an address to return to, this discussion shows that the address is complex, multidimensional, dialectical, relational, and continuously changing. Therefore, it is with this in mind that the self can be the source of a request for authenticity: it is not a transparent, fixed, and unchanging point of reference, but multilayered, opaque, and evolving. If authenticity is unity of the self, it requires the recognition of the self's many nuances and complexities, as well as its exposure to change and transformation, rather than a recipe for rigid resistance to any change. This requires going beyond the battles between those who have tried to turn the human into the centre of the world and those who deny the existence of the human as a conscious, active agent. The emphasis on the embodied materiality of mental processes, on the unconscious and the fragmented structure of the mind, and on the embedded social character of the self, all explicitly challenge the mythology of human omnipotence. This is a battle in which human abilities are either exaggerated or denied. The exaggeration of these abilities has led to numerous crises, of which the climate emergency may be the most significant. Meanwhile denying these abilities ignores the ability to choose between different possibilities, as such a denial would reduce human actions to mere randomness. The possibilities for choice are always limited within the often-unfavourable circumstances, but they do exist in the interstices of events, generating a degree of hope and a space for action. The quest for authenticity as a desire for self-awareness and self-confidence reflects this intrinsic need for hope to have some control over the course of events.

The sense of self is often associated with the mind, which seems to be distinctive from the body; but the mind is an integral part of the whole body and interacts with the environment. Therefore, the self is embodied and embedded, combining biological, psychological, and social processes. As Spinoza argued, the mind is the knowledge of the body, and as the body and the environment change continually, brain's plasticity responds by rewiring itself and changing human subjectivity. For cognitive and neuroscientists, the mind is a process produced by brain–body interaction, consciousness is a workspace for concentrating the mental processes in a centre of attention, and the sense of self is the product of this consciousness being directed inwards. However, this research has not been engaged with explaining the first-person experience of subjectivity, and the relationship between subjectivity and materiality, where the meaning of qualitative concepts such as authenticity lies.

For psychoanalysts, this sense of the self is only the conscious part of the psyche, which also includes an unconscious part, always beyond reach. The psyche is fragmented, a scene of conflict between biological and social forces, between desires and norms, tearing apart what is thought to be the self, leaving only an imagined self, a decentred mirage. This divided self with its unreachable, unknowable, and shifting core, punctured with unbridgeable voids between the speaker and the language, would hardly provide a solid core for the idea of being true to the self. Over the centuries, different theories have analysed the mind into its constituent parts, and some have questioned the existence of a unity between them. The absence or fragmentation of the self would undermine the possibility of searching for an inner point of origin for an authentic self.

Ethical systems have tended to prioritize the social over the individual, to make life in society possible, demanding the individuals to be true to themselves by avoiding false pretensions, while punishing them when their features do not match social expectations. These socio-centred systems were challenged by the modern idea of the self. However, this self is not a detached will in full control of itself and out to conquer the world, to which the European Enlightenment aspired, but part of a social and material world, and formed through being raised in and always in interaction with this world. The self is mediated rather than detached, it is shaped through intersubjective relations, driven by the desire for social recognition. It is formed through reflexive thinking, looking at the self from within and from outside, engaged in a continuous internal dialogue in anticipation of communicating with others, and in the construction of a narrative that links up the fragments of memories and mental processes. The self is the result of an ongoing dialogue between the inside and outside, between the body and environment, and by extension between the inside and outside spaces of the city. If the self is to serve as a point of reference for authenticity, the search for both the self and authenticity should focus on these interstices, as they unfold and transform through time, rather than the imagined pure and fixed interiors.

In the maxim 'be true to your self', if the self is thought to be an interiority that is detached from the outside, the private space of the home would become an expression of authenticity. However, like the inner space of human subjectivity, the home, like all other private spaces, is not a pure and detached space, but one which is constituted by inner diversities and multiple relations with the outside. The search for urban authenticity cannot take us merely to the private inner spaces, but to the city as a whole, in all its diversity and range of concealed and disclosed spaces. If authenticity demands that the outside be a true reflection of the inside, the public environment should be organized like a private space, which is a practical impossibility. If the outside is thought to be inauthentic, then many features of public life and environment, such as facades and ornamentations, or the interpersonal relations and public performances, would have to be rejected as inauthentic. At the same time, if private life in the home is opened to and shared with others, a realm of intimacy is denied to the household.

The core of urban life is not a pure origin located in the private space of the house, as it is not a homogeneous and secluded space, especially with the growth of the new institutions and technologies. Inside and outside spaces are interdependent, rather than detached. The pursuit of authenticity does not mean the prevalence of one over the other, nor the unification of the two, but paying attention to the articulation of their relationship. Being true to the self does not mean drawing harsh boundaries around inner spaces, but to see both inside and outside spaces as interdependent parts of the same environment. The organization of the urban space into inside and outside spaces, the character of the boundaries that separate them, and the range of spaces that fall between them, reflect the character of the urban society, and whether it is at peace with itself or suffering from high inequality, injustice, and severe violence. The self of the city, if such a thing exists, is not limited to its suppressed memories, hidden corners, and private interiors, so that they can be discovered and disclosed, but this self would also include its exterior, common spaces. As such, this self would have to be found in the totality of the city, in all aspects of its unfolding multiplicity.

The word authenticity is closely related to the word author, where the authority for producing something lies, and where authenticity is claimed though tracing a direct link with that authority. Authenticity, however, is distinctive from autonomy, which advocates taking rational control of the self. For some, like the existentialists, the course of life can and should be self-determined: the authenticity of the self comes from authoring it, like creating a unique and original work of art, in contradistinction from the social norms, which induce conformity. This authoring, nevertheless, may be constructing a narrative that links together some life events to produce a series of masks, socially recognizable performances that respond to different situations. At the opposite end of the spectrum, selves are thought to be socially determined, each course of life playing out a version of an overall spirit, and the freedom of choice is no more than an illusion. Meanings are also seen as socially determined: texts may be written by authors, but afterwards the authors disappear from the scene, and the meanings of the texts are created by the readers. An alternative perspective to the dichotomy of self-determination and social determination is to see how actions are framed by their conditions of possibility but without being determined by them, and how meaning is created dialectically rather than categorically.

As historical creations, cities are always changing, and any attempt at finding the author or the first designer of a city, or of the first practices of city making, will be lost in the mist of time and the evolution of the species. After millennia of nameless city making, the emphasis on a single designer emerges alongside the rise of individualism, humanism, and absolute monarchies, creating a myth of origins for the identity of the city. The well-ordered city was to be shaped in perfect geometry, by a single designer and governed by a single ruler, from Fliarete's utopian Sforzinda, to Descartes' vision of a rational city, Haussmann's fully restructured Paris, and Le Crbusier's grandiose plans for entire cities. In more recent references to a single designer, it is the reputation of the designer and the city that are brought together in search of asserting a status in global competitions. The search for a single origin for a city leads to a unity of politics and aesthetics and to the power that lies at the heart of controlling the narratives of authenticity.

However, the complexity of the urban development processes cannot be captured in the cult of the genius and the stories about single designers, as they are not omnipotent actors, but framed by their paradigms, instruments, and conditions of possibility, always engaged in exchanges with their environment. A wide range of political, economic, and cultural forces are at work in the construction of a city, and while some like the financiers and developers may have an upper hand, and a development may carry the name of a single source of authority, it is in fact the result of the work of many who build and inhabit the city. Any claim to authenticity would have to be assessed within this complex context with its hierarchies and diversities.

If the self of the city is diverse and continually changing, and the author of a city is complex, multiple, and historical, perhaps the authenticity of the city can be found in its materiality, which can be studied in its methods of production, materials, quantity, form, content, function, and use. In their aesthetic theories, Adorno and Benjamin argue for a critical materialist interpretation, which requires a shift of focus from the author to the object, from the artist to artwork, so as to avoid the cult of genius. Rather than aestheticizing politics, as the fascists had done, they argue for a politicization of aesthetics. In his concept of aura, Benjamin encapsulates uniqueness, originality, and authenticity, which transfer authority to the object in its presence.

In contrast, rather than presence, Adorno emphasizes the content, so that the work can be examined in its critical relation to society and history, rather than being associated with the authority of the author, which detaches it from its content and context.

These two views may be expressed as presence and representation, which both need critical scrutiny. The question is how the aura is recognized, as it requires some specialist knowledge and sense to distinguish the ordinary from the extraordinary. Another question is whether any work can be outside its society and epoch, thereby posing the problem of representing society and history by an artwork. Any object may have an aura for someone, as the psychologists can show, and any object can represent a society and epoch, as the archaeologists testify. This tends to shift the emphasis away from the object to the diversity of experiences, which could democratize the judgement of authenticity, while raising the problem of the required capacities for understanding and judgement without which the notion of authenticity could lose its meaning.

Two aspects of the artworks, their quantity and quality, that is, their uniqueness and aesthetic value, have been closely associated with authenticity. However, according to the criterion of uniqueness, all objects and life forms are singularities, whether or not considered as artworks. All cities are unique, as no two cities are the same, and all the animate and inanimate elements within them are singularities. Uniqueness and rarity are key indicators of an object's exclusivity and desirability, but they are not sufficient for determining authenticity. The origin of authenticity may also be the material with which something is made, rather than its quantity, so we may have an authentic concrete building, which shows a link between a building and a building material as its criterion of authenticity, although this relationship becomes much weaker in the diversity of building materials in a city.

The origin of authenticity may be searched in the form of an object, its appearance, and what makes it recognizable as what it is. This makes design as the form and composition of objects a primary indicator of aesthetic value and authenticity. Formalism was widely adopted by modernist art and architecture, although in architecture and urbanism, form is not a mere appearance but the order of objects, spaces, and social life. Through revolutionizing form, modernist planning and design emphasize content, and its functionalism brings forward the social character of space, albeit narrowly defined. Nevertheless, as the urban experience has demonstrated, the designer's intended function and content are not necessarily the way a space is used, now or in the future. The primary character of a place may be determined through its constantly changing uses. The materiality of the city, in all these dimensions, shows that authenticity cannot be limited to unique objects, fixed characteristics, single origins, and aesthetic appeals.

The origin of authenticity may be sought in events, the unique occurrences that appear to defy regularity and conformity. However, from a broader perspective, all physical objects and human activities may be considered as unique events, as no two persons or places are the same. The essence of the city could be defined as the absence of an essence, as multiplicity and randomness of events. A flurry of temporary performances is taken to be the real city, considering the outside rather than the inside to be determining what an authentic city is. The vibrant city with its visual culture filled with ephemeral images, fleeting appearances, chaotic scenes, and crowded places, can give the impression that an authentic city is a busy place of random events. However, events do not occur in a void but are related to other events before and after them, and what seems a mere surface or a random event may be part of a pattern of regularity. Moreover, by grouping the unique events under the universal title of authenticity, these events seem to lose their singularity and be absorbed into a general pattern.

Any event can be selected as a point of reference, a turning point that can influence and transform the following events. In urbanism, physical infrastructure and social institutions may be excluded from the list of events, for being distinctive from ephemeral scenes of everyday urban life. While major events may have transformative powers, everyday occurrences also can, through their cumulative effect, lead to major changes. In urbanism, the emphasis on major events has led the attention to main nodes and monuments, and to the display windows of the city, while undermining the significance of small details, ordinary spaces, and excluded corners of the city. Despite their apparent randomness, these small events find a regular pattern in the habits and behaviour of urban dwellers, which may be subject to strict orders and considered to be inauthentic and alienating. Randomness and regularity, events and routines, are interconnected, and when selecting one as the origin of authenticity, the other is inevitably included.

The origin of authenticity may be time, which involves selecting a point in time, in the past or present, a real or imagined rupture in the flow of time that serves as the starting point, as the original source of something. The link with the origin is aimed at showing that an object was rooted in, or can represent, a specific period of history. In the context of accelerated temporality, authenticity expresses a desire for temporal continuity, constructing a bridge to the past and to the future, organizing the fragments of meaning into a coherent storyline. However, in the flow of time, where everything has a past, finding a definite moment as the pure origin of something may not be possible, as those are often moments of transformation and change. Through cultural habits and social conventions, some historical moments are selected, often by those who are powerful enough to shape the cultural agenda, as the points of reference and as narrative bridges to overcome the interrupted memories. The process of authentication constructs a narrative identity, bridging between the object and that temporal point of reference. Temporal authenticity involves making a choice from the overlapping shapes of the historical tapestry. The questions that emerge are about what signifies the past, which past is selected, whose memory is appreciated, whose past is recounted, through what narrative, and how to deal with it.

While the language of authenticity has often been used by antimodern attitudes, it can also describe its nemesis, modernism, even if modernists were blamed for opposing it. In urbanism, this has been manifested by the competing paradigms of modernism, which selected the present as an origin for future authenticity, and conservatism, which selected the past as the source of present authenticity. While one paradigm severed the lines of temporal continuity, thereby causing anxiety and cultural instability, the other tried to mend these lines through protecting and conserving the heritage of the past. Neither claim to authenticity, however, could survive the test of time unscathed. Beyond the temporal point of reference, both paradigms claim to represent a particular period, but each period is filled with multiple and diverse attributes, and history cannot be reduced to single representations or solitary paradigms. The claims of science, technology, and progress were utilized in the colonization of the world, mass destruction, and degradation of the natural habitat. Meanwhile, the claims of cultural heritage were contested through its differential emphasis on some layers of the palimpsest, which reflected power imbalances and historical injustices, and by the forces of commodification that frame the heritage and tourism industries. The expressions of temporal authenticity, from designating a historic town as cultural heritage to designing a new town for future generations, are based on potentially conflictual exchanges between different temporalities and subjectivities, rooted in political, economic, and cultural specificities.

The point of origin for authenticity may be space, identifying a particular place from which something has originated, hence establishing a claim for continuity in one place or a relationship between two locales. Spatial authenticity is always intertwined with temporal authenticity, constructing a point of origin that triggers a desire to return to an imagined, pure, and uninterrupted continuity in the social and natural world, a search for the unity of being that has been lost in the alienation caused by the modern world. Authenticity is sought in the presumed simplicity and innocence of the natural world, away from the compromises and conventions of the human world. However, this natural world has been radically exploited and degraded to the edge of extinction. Authenticity may be sought in returning to a former spacetime called home, but the place and the person, the home, and the self, are not fixed and have not stayed the same, and a return may not be desirable or possible. Place-related authenticity draws attention to quality, particularity, embodiment, and embeddedness, gathering the fragments of life in a locale, in critical response to the universality of abstract, quantitative space. With the combination of power and particularity comes a dialectic of inside and outside, inclusion and exclusion, with far-reaching consequences, ranging from reactionary to emancipatory movements.

In combination with time, space becomes a key point of origin for identity construction, which inevitably draws on both similarity and difference, and all the challenges that this entails. City administrations around the world compete for distinction through iconic structures, while inherently hoping that such distinction would qualify them for inclusion in a club of similar cities. These attempts at distinctiveness rely on commercial signs, rather than the specificities of the urban social life, which would make any city unique anyway. The claims of making authentic places and authentic neighbourhoods are often associated with the images of vibrant streets, which are taken as a sign of commercial success. The many dimensions of the city are reduced to an image, masking some of the underlying processes, whereby the existing unequal social structures are reproduced and enhanced through the processes of placing and distancing, clustering and separating. The discourse of authenticity may refer to a surface, an image, as its proof of authentication, disregarding the many layers of meaning that constitute the particularity of a place.

The object may be the city and the origin may be selected as society, so that authenticity may be secured through a relationship between the particularity of the urban space and the generality of society. Society with its historical meaning of cohesiveness and association, however, would not describe the fragmented urban society of today, whose complexity cannot be reduced to single descriptions and narrow identities. The multiplicity of layers that make up society, as observed by a multiplicity of observers, perspectives, and experiences, generate numerous narrative identities. In the diversity of cultures within any society, qualities such as authenticity would inevitably be diversified. This is not a path to relativism, but rather a critical examination of the claims that may limit authenticity to powerful forces and iconic places, leading to a more inclusive recognition of the diversity of claims to authenticity at all levels. It shows how objects have a history and a context, rather than existing in a void, and how different forms of authenticity may be found in different parts of a city. Rather than a categorical fixity that is static, universal, and timeless, authenticity is a pursuit of particularity in a dialectical process that unfolds in time, a normative process of judgement that reflects the values and preferences of the judges, which therefore need to be always open to critical examination.

The creative potential of authenticity lies partly in its quest for truthfulness, which forms its critical edge and can take cultural and political forms. Authenticity is a narrative that can empower its narrator, even if the story that is told is no more than mere fiction. Authenticity can be an instrument of narcissism, commodification, and reactionary politics, but it can also be an emancipatory weapon for the suppressed minorities and colonized populations to mobilize themselves through their distinctiveness. It is a bridge between perception and conception, and while it can be used to reassert preconceived ideas, it can also open the door to questioning the status quo. It can be a form of resistance against the normative straitjacket of the social order imposed by bureaucratic and technological structures of capitalist society. However, the expression of this resistance in urban life, and in the design and development of cities, requires going beyond the individualist expressions of nonconformity, which can make dissonance a new form of conformity. The resistance requires including the possibility of a public realm in which truthful mutual linkages may be made so that new social orders can be tested and created. Authenticity appears to be a quest for being different, but it cannot abolish the similarities that relate people to one another and make life in society possible. It is through both similarity and difference that identities are forged and life in society can become possible.

# References


Adorno, Theodor, 1973, *The Jargon of Authenticity*, London: Routledge & Kegan Paul.

Adorno, Theodor, 1984, The essay as form, *New German Critique*, 32 (Spring–Summer), pp.151–171.

Adorno, Theodor, 1991, *The Culture Industry: Selected Essays on Mass Culture*, London: Routledge.

Adorno, Theodor, 2002, *Aesthetic Theory*, London: Continuum.

Adorno, Theodor, 2004, *Negative Dialectics*, London: Routledge.

Agamben, Giorgio, 1998, *Homo Sacer: Sovereign Power and Bare Life,* Stanford, CA: Stanford University Press.

Ainsworth, Thomas, 2020, Form vs. matter, in Edward N. Zalta, ed, *The Stanford Encyclopedia of Philosophy,* Summer 2020 edition, https://plato.stanford.edu/archives/sum2020/entries/form-matter, accessed 21.11.2023.

Akhavi, Shahrough, 1998, Overcoming tradition and modernity: The search for Islamic authenticity, by Robert D. Lee, Book review, *International Journal of Middle East Studies*, 30:3, pp.459–462.

Alberti, Leon Battista, 1988, *On the Art of Building in Ten Books,* Cambridge: The MIT Press.

Aldrich, Megan & Jos Hackforth-Jones, 2012, *Art and Authenticity*, London: Lund Humphries.

Anderson, Benedict, 2006, *Imagined Communities: Reflections on the Origin and Spread of Nationalism*, Revised Edition, London: Verso.

Ansari, Sarah, 2020, How statues fared after independence? *The Guardian*, 19 June 2020, https://www.theguardian.com/world/2020/jun/19/how-statues-fared-after-independence, accessed 25.01.2024.

Anwar, Mohamed, 1979, *The Myth of Return: Pakistanis in Britain*, London: Heinemann.

Arendt, Hannah, 1958, *The Human Condition*, Chicago, IL: University of Chicago Press.

Arendt, Hannah, 1969, Introduction, in Walter Benjamin, ed, *Illuminations,* New York: Schocken Books, pp.1–55.

Arendt, Hannah, 1998, *The Human Condition*, 2nd edition, Chicago, IL: Chicago University Press.

Aristotle, 1991, *The Complete Works of Aristotle*, edited by Jonathan Barnes, Princeton, NJ: Princeton University Press.

Austin, John Langshaw, 1962, *How to Do Things with Words*, Oxford: Clarendon Press.

Badiou, Alain, 2005, *Being and Event*, London: Continuum.

Badiou, Alain, 2009, *Logic of Worlds: The Sequel to Being and Event*, London: Continuum.

Badiou, Alain, 2022, *The Immanence of Truths*, London: Bloomsbury Academic.

Barthes, Roland, 1977, *Image, Music, Text*, London: Fontana.

Baudrillard, Jean, 1995, *Simulacra and Simulation*, Ann Arbor: University of Michigan Press.

Baumeister, Roy F., 1999, ed, *The Self in Social Psychology*, Philadelphia, PA: Psychology Press.

BBC, 2017, *How the Babri Mosque Destruction Shaped India*, *BBC News*, https://www.bbc.co.uk/news/world-asia-india-42219773, accessed 25.01.2024.

BBC, 2020, *Edward Colston Statue: Protesters Tear Down Slave Trader Monument*, *BBC News*, https://www.bbc.co.uk/news/uk-52954305, accessed 25.01.2024.

BBC, 2021, *Colombian Anti-Government Protesters Topple Columbus Statue*, https://www.bbc.co.uk/news/world-latin-america-57651833, accessed 25.01.2024.

BBC, 2023, How I rewired my brain in six weeks, *BBC News*, https://www.youtube.com/watch?v=pi9Xvh-Dva4, accessed 12.05.2024.

Bell, Daniel, 2024, Communitarianism, in Edward N. Zalta & Uri Nodelman, eds, *The Stanford Encyclopedia of Philosophy*, Summer 2024 edition, https://plato.stanford.edu/archives/sum2024/entries/communitarianism, accessed 15.01.2025.

Benjamin, Walter, 1969, *Illuminations*, New York: Schocken Books.

Bennett, Maxwell & Peter Hacker, 2007a, Philosophical foundations of neuroscience: The introduction, in Maxwell Bennett, Daniel Dennett, Peter Hacker & John Searle, eds, *Neuroscience & Philosophy: Brain, Mind, & Language*, New York: Columbia University Press, pp.3–13.

Bennett, Maxwell & Peter Hacker, 2007b, Philosophical foundations of neuroscience: An excerpt from Chapter 3, in Maxwell Bennett, Daniel Dennett, Peter Hacker & John Searle, eds, *Neuroscience & Philosophy: Brain, Mind, & Language*, New York: Columbia University Press, pp.15–33.

Bergson, Henri, 2007, *The Creative Mind: An Introduction to Metaphysics*, Mineola, NY: Dover Publications.

Berlucchi, Giovanni & Henry A. Buchtel, 2009, Neuronal plasticity: Historical roots and evolution of meaning, *Journal of Experimental Brain Research*, 192:3, pp.307–319.

Berman, Marshall, 2009, *The Politics of Authenticity*, new edition, London: Verso.

Bermúdez, Jorge L., 2023, *Cognitive Science: An Introduction to the Science of the Mind*, 4th edition, Cambridge: Cambridge University Press.

Beverland, Michael, 2009, *Building Brand Authenticity: 7 Habits of Iconic Brands*, London: Palgrave Macmillan.

Blakely, Edward J. & Michelle G. Snyder, 1999, *Fortress America: Gated Communities in the United States*, Washington, DC: Brookings Institution Press.

Blumer, Herbert, 1969, *Symbolic Interactionism: Perspective and Method,* Englewood Cliffs, NJ: Prentice-Hall.

Bottéro, Jean, 2000, Religion and reasoning in Mesopotamia, in Jean Bottéro, Clarisse Herrenschmidt & Jean-Pierre Vernant, eds, *Ancestor of the West: Writing, Reasoning and Religion in Mesopotamia, Elam and Greece*, Chicago, IL: The University of Chicago Press, pp.1–66.

Bourdieu, Pierre, 1984, *Distinction: A Social Critique of the Judgement of Taste*, Cambridge, MA: Harvard University Press.

Bourdieu, Pierre, 2000, *Pascalian Meditations*, Cambridge: Polity Press.

Bouveresse, Jacque, 1995, *Wittgenstein Reads Freud: The Myth of the Unconscious*, Princeton, NJ: Princeton University Press.

Branaman, Ann, 2001, *Self and Society*, Oxford: Blackwell.

Briggs, Asa, 1968, *Victorian Cities*, Harmondsworth and Middlesex: Penguin.

Bronner, Stephen Eric, 2021, *Existentialism, Authenticity, Solidarity,* London: Routledge.

Brook, Peter, 1990, *The Empty Space*, London: Penguin Books.

Buchanan, Ian, 2010, *A Dictionary of Critical Theory*, Oxford: Oxford University Press.

Burgoyne, Bernard, 2003, From the letter to the matheme: Lacan's scientific methods, in Jean-Michel Rabaté, ed, *The Cambridge Companion to Lacan*, Cambridge: Cambridge University Press, pp.69–85.

Butler, Judith, 1990, *Gender Trouble: Feminism and the Subversion of Identity*, London: Routledge.

Calder, Barnabas, 2016, *Raw Concrete: The Beauty of Brutalism*, London: Heinemann.

Calhoun, Craig, 2002, ed, *Dictionary of Social Sciences*, Oxford: Oxford University Press.

Calvino, Italo, 1997, *Invisible Cities*, London: Vintage.

Cambridge University, 2023, *Eddington (Northwest Cambridge Development)*, https://www.accommodation.cam.ac.uk/FindAHome/UniversityAccommodation/UniversityOwnedProperties/NorthWestCambridge, accessed 8.08.2023.

Caputo, John D., 2018, *Hermeneutics: Facts and Interpretation in the Age of Information*, London: Pelican Books.

Castells, Manuel, 1997, *The Power of Identity*, Oxford: Blackwell.

Chipman, Susan, 2014, ed, *The Oxford Handbook of Cognitive Science*, Oxford: Oxford University Press.

Coleman, Janet, 1996, ed, *The Individual in Political Theory and Practice*, Oxford: Clarendon Press.

Conzen, Michael Robert Gunter, 1960, *Alnwick, Northumberland, a Study in Town-Plan Analysis*, London: IBG.

Cornford, Francis MacDonald, 1976, The invention of space, in M. Čapek, ed, *The Concepts of Space and Time*, Boston Studies in the Philosophy of Science, Vol. XXII, Dordrecht and Boston, MA: D. Reidel Publishing, pp.3–16.

Cornford, Francis MacDonald, 1997, *Plato's Cosmology: The Timaeus of Plato*, Indianapolis, IN: Hackett.

Coupe, Laurence, 2000, *The Green Studies Reader: From Romanticism to Ecocriticism*, London: Routledge.

Coveney, Peter & Roger Highfield, 1991, *The Arrow of Time: The Quest to Solve the Science's Greatest Mystery*, London: Flamingo.

Cranston, Maurice, 1968, Introduction, in Jean-Jacques Rousseau, ed, *The Social Contract*, London: Penguin, pp.9–43.

Cullingworth, John Barry & Roger W. Caves, 2013, *Planning in the USA: Policies, Issues and Processes*, 4th edition, London: Routledge.

Damasio, Antonio, 2004, *Looking for Spinoza*, London: Vintage.

Darwin, Charles, 1871, *The Descent of Man, and Selection in Relation to Sex*, Volume 1, London: John Murray.

David, Marian, 2022, The correspondence theory of truth, in Edward N. Zalta, ed, *The Stanford Encyclopedia of Philosophy*, Summer 2022 edition, https://plato.stanford.edu/archives/sum2022/entries/truth-correspondence, accessed 13.12.2023.

De Cauwer, Stijn, 2021, Ariella Aïsha Azoulay and Georges Didi-Huberman: The persistence of lost worlds, *Journal of Aesthetics & Culture*, 13:1, 1994179, DOI: 10.1080/20004214.2021.1994179.

De Montaigne, Michel, 2004, *The Complete Essays*, London: Penguin.

Debord, Guy, 1994, *The Society of the Spectacle*, New York: Zone Books.

Demir, V. Metin, 2020, Why is habit the hardest problem for Hegel? *Lo Sguardo – Rivista di Filosofia*: 31:II, pp.1–20.

Dennett, Daniel, 2013, *Intuition Pumps and Other Tools for Thinking*, London: Penguin Books.

Dépelteau, François, 2013, Comparing Elias and Bourdieu as relational thinkers, in François Dépelteau & Tatiana Savoia Landini, eds, *Norbert Elias and Social Theory*, New York: Palgrave Macmillan, pp.275–295.

Descartes, René, 1968, *Discourse on Method and the Meditations*, London: Penguin.

Descartes, René, 1973, *The Philosophical Works of Descartes*, Volume 1, translated by Elizabeth S. Haldane and George Robert Thomson Ross, Cambridge: Cambridge University Press.

Descombes, Vincent, 1995, Foreword, in Jacque Bouveresse, ed, *Wittgenstein Reads Freud*, Princeton, NJ: Princeton University Press, pp.vii–xiii.

Dewey, John, 1946 [1927], *The Public and its Problems: An Essay in Political Inquiry*, Chicago, IL: Gateway Books.

Dudley, Nigel, 2011, *Authenticity in Nature*, London: Earthscan.

Durkheim, Emile, 1982, *The Rules of Sociological Method*, New York: The Free Press.

Economist, 2024, God of the Greek kitchen, *The Economist*, 30 March 2024, p.77.

Elias, Norbert, 1978, *The Civilizing Process*, Oxford: Blackwell.

Engels, Friedrich, 1993, *The Condition of the Working Class in England*, Oxford: Oxford University Press.

Etzioni, Amitai, 1995, ed, *The Spirit of Community: Rights, Responsibilities and the Communitarian Agenda*, London: Fontana Press.

Falahat, Somaiyeh & Ali Madanipour, 2019, Lifeworld and social space: Spatial restructuring and urban governance in Berlin, *disP: The Planning Review*, 55:4, pp.46–59.

Farrer, Martin, 2020, Who was Edward Colston and why was his Bristol statue toppled? *The Guardian*, 8 June 2020, https://www.theguardian.com/uk-news/2020/jun/08/who-was-edward-colston-and-why-was-his-bristol-statue-toppled-slave-trader-black-lives-matter-protests, accessed 28.01.2024.

Feke, Jacqueline, 2018, *Ptolemy's Philosophy: Mathematics as a Way of Life*, Princeton, NJ: Princeton University Press.

Ferrara, Alessandro, 1998, *Reflective Authenticity: Rethinking the Project of Modernity*, London: Routledge.

Filarete, 1965, *Treatise on Architecture*, New Haven, CT: Yale University Press.

Finney, Patrick, 2019, *Authenticity: Reading, Remembering, Performing*, London: Routledge.

Foster, Peter, 2023, Hopes of turning Cambridge into 'science capital of Europe' collide with reality, *Financial Times*, 8 August.

Foucault, Michel, 1981, The order of discourse, in Robert Young, ed, *Untying the Text: A Poststructuralist Reader*, London: Routledge and Kegan Paul, pp.51–78.

Foucault, Michel, 1986, *The Care of the Self*, New York: Pantheon Books.

Foucault, Michel, 1988, Technologies of the self, in Luther Martin, Huck Gutman & Patrick Hutton, eds, *The Technologies of the Self: A Seminar with Michel Foucault*, London: Tavistock publications, pp.16–49.

Frankopan, Peter, 2015, *The Silk Roads: A New History of the World*, London: Bloomsbury.

Freud, Sigmund, 1962, *The Ego and the Id*, London: Hogarth Press and the Institute of Psycho-Analysis.

Freud, Sigmund, 1991, *Civilization, Society and Religion*, Volume 12, Penguin Freud Library, London: Penguin Books.

Freud, Sigmund, 2008, *The Interpretation of Dreams*, Oxford: Oxford University Press.

Gadamer, Hans-Georg, 1976, *Philosophical Hermeneutics*, Berkeley: University of California Press.

Gadamer, Hans-Georg, 2004, *Truth and Method*, London: Continuum.

Gamel, Mary-Kay, 2013, Can 'democratic' modern stagings of ancient drama be 'authentic'? in Lorna Hardwick & Stephen Harrison, eds, *Classics in the Modern World: A Democratic Turn?* Oxford: Oxford University Press, pp.182–195.

Gans, Herbert, 1968, *People and Plans*, New York: Basic Books.

Garratt, Lindsey, 2018, *The Body, Authenticity and Racism*, London: Routledge.

Gellner, Ernest, 1983, *Nations and Nationalism*, Oxford: Blackwell.

George, Bill, Herminia Ibarra, Rob Goffee & Gareth Jones, 2017, *Authentic Leadership*, Brighton, MA: Harvard Business Press.

Giedion, Sigfried, 1967, *Space, Time and Architecture: The Growth of a Tradition*, 5th edition, Cambridge, MA: Harvard University Press.

Gilovich, Thomas, Dacher Keltner, Serena Chen & Richard Nisbett, 2016, *Social Psychology*, 4th edition, New York: W.W. Norton and Company.

Goffman, Erving, 1969, *The Presentation of Self in Everyday Life*, London: Allen Lane, The Penguin Press.

Golomb, Jacob, 1995, *In Search of Authenticity: From Kierkegaard to Camus*, London: Routledge.

Goodall, Brian, 1987, The Penguin Dictionary of Human Geography, Harmondsworth: Penguin.

Gray, Jeremy, 1989, *Ideas of Space: Euclidean, Non-Euclidean and Relativistic*, 2nd edition, Oxford: Clarendon Press.

Green, Mitchell, 2021, Speech acts, in Edward N. Zalta, ed, *The Stanford Encyclopedia of Philosophy*, Fall 2021 edition, https://plato.stanford.edu/archives/fall2021/entries/speech-acts, accessed 15.01.2025.

Greenfield, Susan, 2000, *The Private Life of the Brain*, London: Allen Lane, The Penguin Press.

Habermas, Jürgen, 1984, *The Theory of Communicative Action*, London: Heinemann.

Hadot, Pierre, 1995, *Philosophy as a Way of Life*, Oxford: Blackwell.

Hanly, Charles, 1990, The concept of truth in psychoanalysis, *International Journal of Psychoanalysis*, 71:Pt 3, pp.375–383.

Hannah, Dorit, 2019, *Event-Space: Theatre Architecture and the Historical Avant-Garde*, London: Routledge.

Harvey, David, 2005, *Paris: The Capital of Modernity*, London: Routledge.

Hegel, Georg Wilhelm Friedrich, 1975, *Aesthetics: Lectures on Fine Art*, Volume 1, Oxford: Clarendon Press.

Hegel, Georg Wilhelm Friedrich, 1991, *Elements of the Philosophy of Right*, Cambridge: Cambridge University Press.
Hegel, Georg Wilhelm Friedrich, 2001, *Philosophy of Right*, Kitchener, ON: Batoche Books.
Heidegger, Martin, 1962, *Being and Time*, Oxford: Blackwell.
Heidegger, Martin, 2011, *Basic Writings*, edited by David Farrell Krell, London: Routledge.
Heyden, Tom, 2015, The 10 greatest controversies of Winston Churchill's career, *BBC*, 26 January, https://www.bbc.co.uk/news/magazine-29701767, accessed 28.01.2024.
Hobbes, Thomas, 1929, *Leviathan*, Oxford: Clarendon Press.
Hobsbawm, E. J., 1992, *Nations and Nationalism Since 1780*, 2nd Edition, Cambridge: Cambridge University Press.
Hogg, Michael & Graham Vaughn, 2018, *Social Psychology*, 8th edition, Harlow: Pearson.
Hollis, Martin, 2003, *The Philosophy of Social Science*, Cambridge: Cambridge University Press.
Homer, 2018, *Odyssey*, translated by Peter Green, Oakland, CA: University of California Press.
Howard, Ebenezer, 2003, *To-Morrow: A Peaceful Path to Real Reform*, London: Routledge.
Hume, David, 1826, *The Philosophical Works of David Hume*, Edinburgh and London: Adam Black & William Tait.
ICCROM, 2000, *Riga Charter on Authenticity and Historical Reconstruction in Relationship to Cultural Heritage*, https://www.iccrom.org/sites/default/files/publications/2020-05/convern8_07_rigacharter_ing.pdf, accessed 22.01.2024.
ICOMOS, 1965, *The Venice Charter*, The Venice Charter – International Council on Monuments and Sites (icomos.org), accessed 20.01.2024.
Iversen, Margaret & Stephen Melville, 2010, *Writing Art History: Disciplinary Departures*, Chicago, IL: The University of Chicago Press.
Jacobs, Jane, 1961, *The Death and Life of Great American Cities*, New York: Vintage Books.
Jencks, Charles, 1992, ed, *The Post-Modern Reader*, London: Academy Editions.
Jenkins, Richard, 1996, *Social Identity*, London: Routledge.
Joas, Hans & Wolfgang Knöbl, 2009, *Social Theory: Twenty Introductory Lectures*, Cambridge: Cambridge University Press.
Johnston, Adrian, 2024, Jacques Lacan, in Edward N. Zalta & Uri Nodelman, eds, *The Stanford Encyclopedia of Philosophy*, Summer 2024 edition, https://plato.stanford.edu/archives/sum2024/entries/lacan, accessed 15.01.2025.
Joseph, Stephen, 2016, *Authentic: How to Be Yourself and Why It Matters*, London: Hachette.
Jung, Carl, 2014, *The Undiscovered Self*, London: Routledge.
Kant, Immanuel, 1987, *Critique of Judgment*, Indianapolis, IN: Hackett.
Kant, Immanuel, 1995, What is enlightenment? in Isaac Kramnick, ed, *The Portable Enlightenment Reader*, London: Penguin, pp.1–7.
Kisner, Manja & Jörg Noller, 2020, eds, *The Concept of Will in Classical German Philosophy*, Berlin: De Gruyter.
Kohl, Christoph, Barbara Christophe, Heike Liebau & Achim Saupe, 2021, eds, *The Politics of Authenticity and Populist Discourses*, London: Palgrave Macmillan.

Krauthammer, Charles, 1999, Person of the century, *Washington Post*, Friday, 31 December, p.A31.
Lacan, Jacques, 2006, *Écrits,* New York: W.W.Norton.
Lau, Hakwan, 2022, *In Consciousness We Trust: The Cognitive Neuroscience of Subjective Experience*, New York: Oxford University Press.
Laugier, Marc-Antoine, 1977, *An Essay on Architecture*, Los Angeles, CA: Hennessey & Ingalls Inc.
Le Corbusier, 1986, *Towards a New Architecture*, Mineola, NY: Dover Publications.
Le Corbusier, 1987, *The City of To-Morrow and its Planning*, Mineola, NY: Dover Publications.
Lee, Robert, 1997, *Overcoming Tradition and Modernity: The Search for Islamic Authenticity*, Boulder, CO: Westview Press.
Lefebvre, Henri, 1991, *The Production of Space*, Oxford: Blackwell.
LeGates, Richard & Fred Stout, 2020, eds, *The City Reader*, 7th edition, New York: Routledge.
Leibniz, Gottfried Wilhelm., 1979, The relational theory of space and time, in John Jamieson Carswell Smart, ed, *Problems of Space and Time*, New York: Macmillan, pp.89–98.
Lindholm, Charles, 2007, *Culture and Authenticity*, Hoboken, NJ: Wiley-Blackwell.
Lippmann, Walter, 1993, *The Phantom Public*, New Brunswick, NJ: Transaction Publishers.
Llewlyn-Davies, Lord, 1972, Changing goals in design: The Milton Keynes example, in E. Evans, ed, *New Towns: The British Experience*, London: Charles Knight & Co., pp.102–116.
Llewlyn-Davies, Weeks & Partners, 1966, *Washington New Town, Master Plan and Report*, Washington, DC: Tyne and Wear; Washington Development Corporation.
Locke, John, 1821, *Two Treatises of Government*, London: Whitmore and Fenn, and C. Brown.
Lovell, Jane & Chris Bull, 2019, *Authentic and Inauthentic Places in Tourism: From Heritage Sites to Theme Parks,* London: Routledge.
Lovell, Jane & Sam Hitchmough, eds, 2020, *Authenticity in North America: Place, Tourism, Heritage, Culture and the Popular Imagination,* London: Routledge.
Luepnitz, Deborah, 2003, Beyond the phallus: Lacan and feminism, in Jean-Michel Rabaté, ed, *The Cambridge Companion to Lacan*, Cambridge: Cambridge University Press, pp.221–237.
Lynch, Kevin, 1960, *The Image of the City*, Cambridge: MIT Press.
Machiavelli, Niccolò, 1950, *The Discourses of Niccolò Machiavelli*, London: Routledge and Paul.
Madanipour, Ali, 1996, *Design of Urban Space: An Inquiry into a Socio-Spatial Process*, Chichester: John Wiley.
Madanipour, Ali, 1997, Ambiguities of urban design, *Town Planning Review*, 68:3 (July), pp.363–383.
Madanipour, Ali, 2001, How relevant is 'planning by neighbourhoods' today? *Town Planning Review,* 72:2, pp.171–191.
Madanipour, Ali, 2003, *Public and Private Spaces of the City*, London: Routledge.
Madanipour, Ali, 2006, Roles and challenges of urban design, *Journal of Urban Design*,11:2, pp.173–193.
Madanipour, Ali, 2007, *Designing the City of Reason: Foundations and Frameworks*, London: Routledge.

Madanipour, Ali, 2010, ed, *Whose Public Space? International Case Studies in Urban Design and Development*, London: Routledge.

Madanipour, Ali, 2011, *Knowledge Economy and the City: Spaces of Knowledge*, London: Routledge.

Madanipour, Ali, 2012, Neighbourhood governance, in Susan J. Smith, Marja Elsinga, Lorna Fox O'Mahony, Ong Seow Eng, Susan Wachter & Montserrat Pareja Eastaway, eds, *International Encyclopedia of Housing and Home*, Volume 5, Oxford: Elsevier, pp.61–66.

Madanipour, Ali, 2013, The identity of the city, in Silvia Serreli, ed, *City Project and Public Space,* Dodrecht: Springer, pp.49–63.

Madanipour, Ali, 2014, *Urban Design, Space and Society,* Basingstoke: Palgrave Macmillan.

Madanipour, Ali, 2015, ed, *Planning Theory*, Four volumes, Critical Concepts in Built Environment Series, London: Routledge.

Madanipour, Ali, 2016, Culture and tolerance in public space, in Stefano Moroni & David Weberman, eds, *Space and Pluralism: Can Contemporary Cities Be Places of Tolerance?* Budapest: Central European University Press, pp.35–54.

Madanipour, Ali, 2017, *Cities in Time: Temporary Urbanism and the Future of the City*, London: Bloomsbury.

Madanipour, Ali, 2020, Social exclusion, space and time, in Richard LeGates & Fred Stout, eds, *The City Reader,* 7th edition, New York: Routledge, pp.206–216.

Madanipour, Ali, 2021, Beyond placing and distancing: Public space for inclusive cities, *Harvard Design Magazine*: 49, pp.74–79.

Madanipour, Ali, 2022a, Critical reflections on care, in Angelika Gabauer, Sabine Knierbein, Nir Cohen, Henrik Lebuhn, Kim Trogal, Tihomir Viderman & Tigran Haas, eds, *Care and the City: Encounters with Urban Studies*, London: Routledge, pp.15–23.

Madanipour, Ali, 2022b, Filarete, in Marco Sgarbi, ed, *Encyclopedia of Renaissance Philosophy*, Switzerland: Springer, pp.1235–1237.

Madanipour, Ali, 2023, *Rethinking Public Space,* Cheltenham: Edward Elgar.

Madanipour, Ali, forthcoming, Revisualising cities and cityscapes, in Barry Sandwell & Martin Hand, eds, *The Bloomsbury Emcyclopedia of Visual Culture*, London: Bloomsbury.

Madanipour, Ali, Angela Hull & Patsy Healey, 2017, eds, *The Governance of Place: Space and Planning Processes,* London: Routledge.

Malabou, Catherine, 2010, *Plasticity at the Dusk of Writing*, New York: Columbia University Press.

Marcuse, Herbert, 1948, Existentialism: Remarks on Jean-Paul Sartre's L'Etre et le Neant, *Philosophy and Phenomenological Research*, 8:3 (March), pp.309–336.

Markosian, Ned, 2014, Time, in Edward N. Zalta, ed, *The Stanford Encyclopedia of Philosophy*, Spring 2014 edition, https://plato.stanford.edu/archives/spr2014/entries/time, accessed 1.09.2015.

Martin, Rux, 1988, Truth, power and self: An interview with Michel Foucault, in Luther Martin, Huck Gutman & Patrick Hutton, eds, *The Technologies of the Self: A Seminar with Michel Foucault*, October 25, London: Tavistock Publications, pp.9–15.

Marx, Karl & Friedrich Engels, 1985, *The Communist Manifesto*, London: Penguin.

Mayhew, Susan & Penny Anne, 1992, *The Concise Oxford Dictionary of Geography*, Oxford: Oxford University Press.

McKusick, James, 2000, *Green Writing: Romanticism and Ecology*, London: Macmillan.
Mendelson, Jack, 1979, The Habermas-Gadamer Debate, *New German Critique*: 18 (Autumn), pp.44–73.
Miao Xu & Zhen Yang, 2009, Design history of China's gated cities and neighbourhoods: Prototype and evolution, *Urban Design International*, 14:2, pp.99–117.
Micheau, Françoise, 2015, Baghdad, an imperial foundation (762–836CE), in Norman Yoffee, ed, *The Cambridge World History*, Cambridge: Cambridge University Press, pp.397–415.
Midgley, Mary, 2018, *What Is Philosophy For?* London: Bloomsbury.
Morris, Anthony Edwin James, 1979, *History of Urban Form*, London: George Godwin.
Morris, William Edward & Charlotte R. Brown, 2022, David Hume, in Edward N. Zalta, ed, *The Stanford Encyclopedia of Philosophy*, Summer 2022 edition, https://plato.stanford.edu/archives/sum2022/entries/hume, accessed 30.11.2023.
Mumford, Lewis, 1954, The neighbourhood unit, *Town Planning Review*, 24, pp.256–270.
Nagel, Thomas, 1998, Concealment and exposure, *Philosophy and Public Affairs*, 27:1 (Winter), pp.3–30.
Newton, Isaac, 1845, *The Mathematical Principles of Natural Philosophy*, translated by Andrew Motte, New York: Daniel Adee.
Nietzsche, Friedrich, 2001, *The Gay Science*, Cambridge: Cambridge University Press.
Nietzsche, Friedrich, 2005, *Thus Spoke Zarathustra*, Oxford: Oxford University Press.
Nietzsche, Friedrich, 2007, *Ecce Homo*, Oxford: Oxford University Press.
Nobus, Dany, 2003, Lacan's science of the subject: Between linguistics and topology, in Jean-Michel Rabaté, ed, *The Cambridge Companion to Lacan*, Cambridge: Cambridge University Press, pp.50–68.
Norberg-Schultz, Christian, 1980, *Meaning in Western Architecture*, London: Studio Vista.
Odegaard, Nancy & Vicki Cassman, 2020, Authentication and conservation in archaeological science, in Claire Smith, ed, *Encyclopedia of Global Archaeology*, Cham: Springer Nature, pp.1228–1237.
Olson, Eric T., 2023, Personal identity, in Edward N. Zalta & Uri Nodelman, eds, *The Stanford Encyclopedia of Philosophy*, Fall 2023 edition, https://plato.stanford.edu/archives/fall2023/entries/identity-personal/, accessed 29.01.2024.
Osborn, Frederic James & Arnold Whittick, 1963, *The New Towns, the Answer to Megalopolis*, 2nd edition, London: Leonard Hill.
Parliament, 2021, *The Conduct of Lord Mann, Appendix 3: Emails Posted to the Google Group of the Professional Membership Body*, https://publications.parliament.uk/pa/ld5802/ldselect/ldcond/38/3810.htm, accessed 28.01.2024.
Paulle, Bowen, Bart van Heerikhuizen & Mustafa Emirbayer, 2012, Elias and Bourdieu, *Journal of Classical Sociology*, 12:1, pp.69–93.
Pazzanese, Christina, 2020, Ethical concerns mount as AI takes bigger decision-making role in more industries, *Harvard Gazette*, 26 October 2020, https://news.harvard.edu/gazette/story/2020/10/ethical-concerns-mount-as-ai-takes-bigger-decision-making-role/, accessed 11.12.2023.
Perloff, Harvey, 1956, Education of city planners: Past, present and future, *Journal of the American Institute of Planners*, 22:4, pp.186–217.

Pevsner, Nikolaus, 1963, *An Outline of European Architecture*, 7th edition, Harmondsworth: Penguin Books.

Pfaff, Donald, Nora Volkow & John Rubenstein, 2022, eds, *Neuroscience in the 21st Century from Basic to Clinical*, 3rd edition, London: Springer Nature.

Plato, 1997, *Complete Works,* edited by John M. Cooper & Dana S. Hutchinson, Indianapolis, IN: Hackett Publishing Company.

Pluhar, Werner, 1987, Translator's introduction, in Immanuel Kant, ed, *Critique of Judgement*, Indianapolis, IN: Hackett, pp.xxiii–cix.

Potter, David Stone, 2009, *Rome in the Ancient World: From Romulus to Justinian*, London: Thames and Hudson.

Rabaté, Jean-Michel, 2003, ed, *The Cambridge Companion to Lacan*, Cambridge: Cambridge University Press.

Rahula, Walpola, 1974, *What the Buddha Taught*, New York: Green Press.

Rancière, Jacques, 2013, *The Politics of Aesthetics: The Distribution of the Sensible*, London: Bloomsbury.

Reinhoudt, Jurgen & Serge Audier, 2018, *The Walter Lippmann Colloquium: The Birth of Neo-Liberalism*, London: Palgrave-Macmillan.

Ricoeur, Paul, 1991, Narrative identity, *Philosophy Today*, 35:1 (Spring), pp.73–81.

Ricoeur, Paul, 1995, *Oneself as Another*, Chicago, IL: Chicago University Press.

Ricoeur, Paul, 2005, *The Course of Recognition*, Cambridge, MA: Harvard University Press.

Ricoeur, Paul, 2006, *Memory, History, Forgetting,* Chicago, IL: The University of Chicago Press.

Ritchie, Hannah, 2024a, *Captain Cook Statue Vandalised in Melbourne on Eve of Australia Day*, https://www.bbc.co.uk/news/world-australia-68090094, accessed 25.01.2024.

Ritchie, Hannah, 2024b, *Australia Day: The 'Quiet Rebranding' of a Controversial National Holiday*, https://www.bbc.co.uk/news/world-australia-68014555, accessed 29.01.2024.

Robinson, Daniel, 2007, Still looking: Science and philosophy in pursuit of prince reason, in Maxwell Bennett, Daniel Dennett, Peter Hacker & John Searle, eds, *Neuroscience & Philosophy: Brain, Mind, & Language*, New York: Columbia University Press, pp.171–193.

Rock, Paul, 1979, *The Making of Symbolic Interactionism*, London: Palgrave Macmillan.

Rodriguez-Pereyra, Gonzalo, 2019, Nominalism in Metaphysics, in Edward N. Zalta, ed, *The Stanford Encyclopedia of Philosophy* (Summer 2019 edition), https://plato.stanford.edu/archives/sum2019/entries/nominalism-metaphysics/, accessed 17.10.2020.

Rousseau, Jean-Jacque, 1783, *The Confessions*, London: J. Bew.

Rousseau, Nathan, 2002, ed, *Self, Symbols, and Society: Classic Readings in Social Psychology*, Oxford: Rowman & Littlefield.

Rudofsky, Bernard, 1964, *Architecture without Architects*, New York: The Museum of Modern Art.

Rumi, Jalal al-Din Muhammad Balkhi, 1983, *Masnavi Manavi*, Tehran: Amir Kabir.

Said, Edward, 1988, Identity, negation and violence, *New Left Review*: 171 (September–October), pp.46–60.

Sartre, Jean-Paul, 2003, *Being and Nothingness: An Essay on Phenomenological Ontology*, London: Routledge.

Schulze, Daniel, 2017, *Authenticity in Contemporary Theatre and Performance: Make it Real*, London: Bloomsbury.
Schumpeter, Joseph, 2003, *Capitalism, Socialism and Democracy*, London: Taylor and Francis.
Searle, John, 1995, *The Construction of Social Reality*, London: Penguin Books.
Searle, John, 1997, *The Mystery of Consciousness*, London: Granta.
Searle, John, 1999, *Mind, Language and Society*, London: Wiedenfeld & Nicolson.
Seligman, Martin, 2002, *Authentic Happiness: Using the New Positive Psychology to Realize Your Potential for Lasting Fulfillment*, London: Random House.
Sert, Jose Luis, 1944, *Can Our Cities Survive? An ABC of Urban Problems, Their Analysis, Their Solution*, Cambridge, MA: Harvard University Press.
Siep, Ludwig, 2014, *Hegel's Phenomenology of Spirit*, Cambridge: Cambridge University Press.
Simmel, Georg, 1950, The metropolis and mental life, in Kurt Wolff, ed, *The Sociology of Georg Simmel*, New York: The Free Press, pp.409–424.
Sitte, Camillo, 1986, City planning according to artistic principles, in George Collins & Christiane Collins, eds, *Camillo Sitte: The Birth of Modern City Planning*, New York: Rizzoli, pp.129–332.
Slowik, Edward, 2014, Descartes' Physics, in Edward N. Zalta, ed, *The Stanford Encyclopedia of Philosophy*, Summer 2014 edition, https://plato.stanford.edu/archives/sum2014/entries/descartes-physics, accessed 1.09.2015.
Smith, Eliot, Diane Mackie & Heather Claypool, 2015, *Social Psychology*, 4th edition, New York: Psychology Press.
Soler, Colette, 2003, The paradoxes of the symptom in psychoanalysis, in Jean-Michel Rabaté, ed, *The Cambridge Companion to Lacan*, Cambridge: Cambridge University Press, pp.86–101.
Spinoza, Benedict de, 1996, *Ethics*, London: Penguin Books.
Stovel, Herb, 2008, Origins and influence of the Nara document on authenticity, *APT Bulletin*, 39:2/3, pp.9–17.
Sullivan, Louis, 1896, The tall office building artistically considered, *Lippincott's*, pp.403–409.
Tabari, Muhammad, 1995, *The History of al-Tabari: Abbasid Authority Affirmed*, Albany: State University of New York Press.
Tate, Laura & Brettany Shannon, 2019, eds, *Planning for AuthentiCITIES*, London: Routledge.
Taylor, Charles, 1989, *Sources of the Self: The Making of the Modern Identity*, Cambridge: Cambridge University Press.
Taylor, Charles, 1991, *The Ethics of Authenticity*, Cambridge, MA: Harvard University Press.
Thacker, Karissa, 2016, *The Art of Authenticity: Tools to Become an Authentic Leader and Your Best Self*, Hoboken, NJ: Wiley.
Thomas, Keith, 2020, *In Pursuit of Civility: Manners and Civilization in Early Modern England*, New Haven, CT: Yale University Press.
Thorne, Kip, 2013, *What Is Space-Time?* https://www.youtube.com/watch?v=mvdlN4H4T54, accessed 26.10.2021.
Tillich, Paul, 1944, Existential philosophy, *Journal of the History of Ideas*, 5:1 (January), pp.44–70.
Tönnies, Ferdinand, 1957, *Community and Society,* New York: Harper and Row.
Trilling, Lionel, 1972, *Sincerity and Authenticity*, London: Oxford University Press.

Tuan, Yi Fu, 1977, *Space and Place: The Perspective of Experience*, London: Edward Arnold.

Umbach, Maiken & Mathew Humphrey, 2018, *Authenticity: The Cultural History of a Political Concept*, London: Palgrave Macmillan.

UNESCO, 1994, *The Nara Document on Authenticity*, https://whc.unesco.org/archive/nara94.htm, accessed 20.01.2024.

UNESCO, 2002, Resolutions, Records of the General Conference, 31st session, Volume 1, Paris: United Nations Educational, Scientific and Cultural Organization.

UNESCO, 2017, *Operational Guidelines for the Implementation of the World Heritage Convention*, Paris: World Heritage Centre, https://whc.unesco.org/en/guidelines, accessed 21.01.2024.

UNESCO, 2023, *Artificial Intelligence: Examples of Ethical Dilemmas*, 21 April, https://www.unesco.org/en/artificial-intelligence/recommendation-ethics/cases, accessed 11.12.2023.

UNESCO, 2024, *Authenticity*, Paris: World Heritage Centre, https://whc.unesco.org/en/faq/187, accessed 21.01.2024.

Urban Task Force, 1999, *Towards an Urban Renaissance*, London: E&FN Spon.

Vannini, Phillip & J. Patrick Williams, 2016, eds, *Authenticity in Culture, Self, and Society*, London: Routledge.

Varga, Somogy & Charles Guignon, 2023, Authenticity, in Edward N. Zalta & Uri Nodelman, eds, *The Stanford Encyclopedia of Philosophy*, Summer 2023 edition, Stanford: Stanford University. https://plato.stanford.edu/archives/sum2023/entries/authenticity/

Walker, Michael, 2019, The importance of an authentic theatre, *Theatre Art Life*, https://www.theatreartlife.com/artistic/authentic-theatre/, accessed 21.02.2024.

Walmsley, Ben, 2013, Co-creating theatre: Authentic engagement or interlegitimation? *Cultural Trends*, 22:2, pp.108–118.

Walzer, Michael, 2021, Individual and community, *Philosophy and Social Criticism*, 47:4, pp.402–407.

Ward, Colin, 1978, *The Child in the City*, London: Architectural Press.

Weiler, Katharina & Niels Gutschow, 2017, eds, *Authenticity in Architectural Heritage Conservation*, Cham: Springer.

Whitrow, Gerald James, 1972, *What Is Time?* London: Thames and Hudson.

Williams, J. Patrick & Kaylan C. Schwarz, 2021, eds, *Studies on the Social Construction of Identity and Authenticity*, London: Routledge.

Williams, Raymond, 1981, *Culture*, Glasgow: Collins.

Williams, Raymond, 1985, *Keywords: A Vocabulary of Culture and Society*, Oxford: Oxford University Press.

Wolters, Albert, 2000, A semantic study of αὐθέντης and its derivatives, *Journal of Greco-Roman Christianity and Judaism*, 1, pp.145–175.

Zizek, Slavoj, 2006, *How to Read Lacan*, London: Granta Books.

Zizek, Slavoj, 2014, *Event: Philosophy in Transit*, London: Penguin.

Zuidervaart, Lambert, 2004, *Aesthetic Truth*, Cambridge: Cambridge University Press.

Zuidervaart, Lambert, 2007, *Social Philosophy after Adorno*, Cambridge: Cambridge University Press.

# Index

Note: *Italic* page numbers refer to figures.

FSC
www.fsc.org
MIX
Papier | Fördert
gute Waldnutzung
FSC® C083411

Kolibri 360 GmbH
Ferdinand-Jühlke-Straße 7
99095 Erfurt, Deutschland
produktsicherheit@kolibri360.de